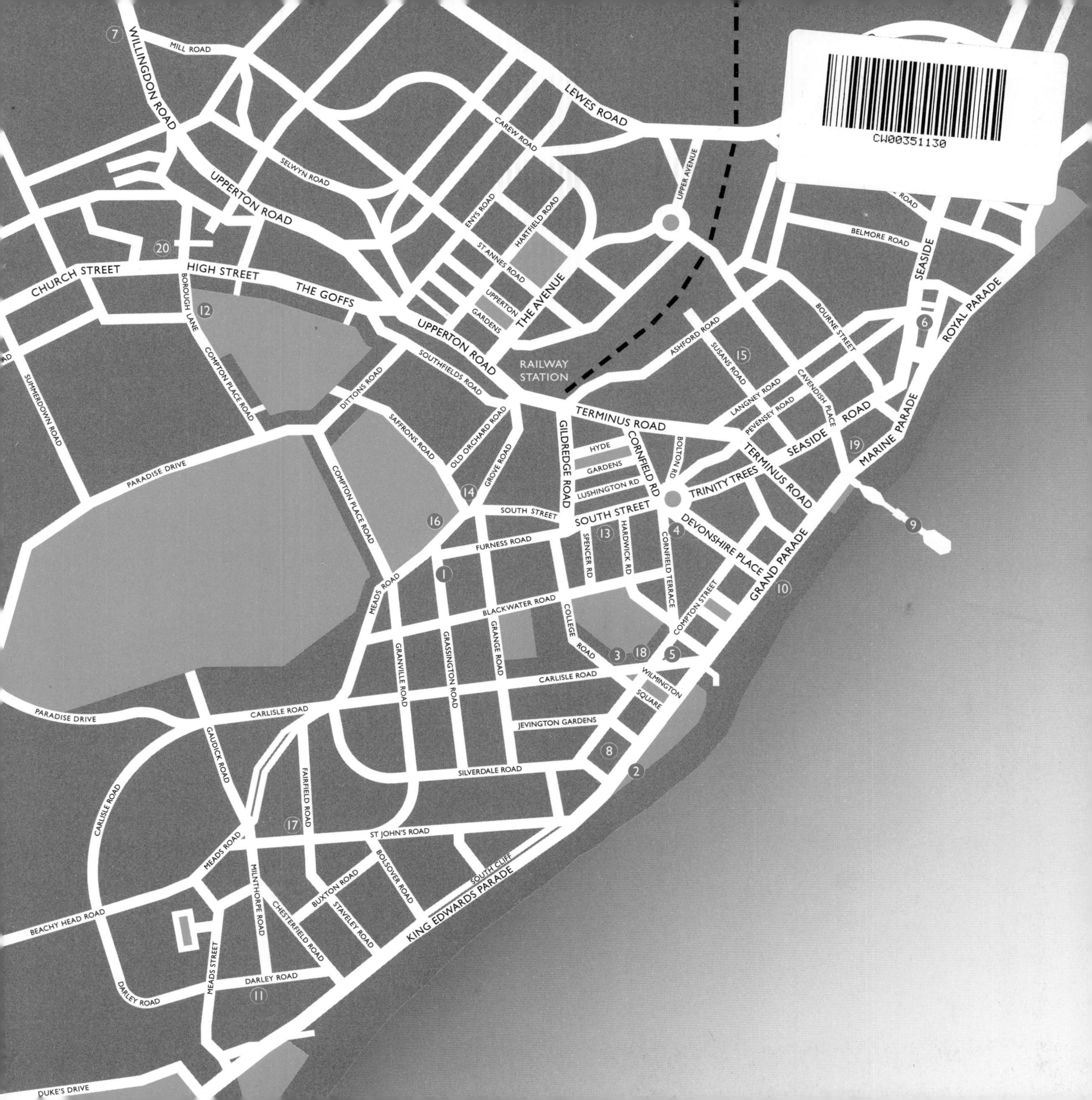

CW00351130
WILLINGDON ROAD
MILL ROAD
LEWES ROAD
CAREW ROAD
SELWYN ROAD
UPPERTON ROAD
UPPER AVENUE
ENYS ROAD
HARTFIELD ROAD
ST ANNES ROAD
BELMORE ROAD
SEASIDE
CHURCH STREET
HIGH STREET
THE GOFFS
BOROUGH LANE
UPPERTON GARDENS
THE AVENUE
UPPERTON ROAD
RAILWAY STATION
ASHFORD ROAD
SUSANS ROAD
BOURNE STREET
ROYAL PARADE
COMPTON PLACE ROAD
SUMMERDOWN ROAD
SOUTHFIELDS ROAD
DITTONS ROAD
SAFFRONS ROAD
OLD ORCHARD ROAD
TERMINUS ROAD
GILDREDGE ROAD
CORNFIELD RD
HYDE GARDENS
LUSHINGTON RD
BOLTON RD
LANGNEY ROAD
PEVENSEY ROAD
CAVENDISH PLACE
SEASIDE ROAD
MARINE PARADE
TERMINUS ROAD
TRINITY TREES
PARADISE DRIVE
GROVE ROAD
SOUTH STREET
SOUTH STREET
COMPTON PLACE ROAD
SPENCER RD
HARDWICK RD
CORNFIELD TERRACE
DEVONSHIRE PLACE
GRAND PARADE
FURNESS ROAD
MEADS ROAD
COMPTON STREET
BLACKWATER ROAD
COLLEGE ROAD
GRANVILLE ROAD
GRASSINGTON ROAD
GRANGE ROAD
CARLISLE ROAD
WILMINGTON SQUARE
PARADISE DRIVE
CARLISLE ROAD
JEVINGTON GARDENS
GAUDICK ROAD
FAIRFIELD ROAD
SILVERDALE ROAD
CARLISLE ROAD
ST JOHN'S ROAD
MEADS ROAD
MILNTHORPE ROAD
BOLSOVER ROAD
BUXTON ROAD
SOUTH CLIFF
KING EDWARDS PARADE
CHESTERFIELD ROAD
STAVELEY ROAD
BEACHY HEAD ROAD
MEADS STREET
DARLEY ROAD
DARLEY ROAD
DUKE'S DRIVE
1
2
3
4
5
6
7
8
9
10
11
12
13
14
15
16
17
18
19
20

eastbourne in detail

First published in Great Britain by DG&P – an imprint of Aves Press Limited,
19 Bolsover Road, Eastbourne, East Sussex BN20 7JG
www.avespress.com

DG&P

Copyright 2015 © Aves Press Limited
ISBN 978-0-9568611-3-9

Printed and bound in Croatia by Imago, Thame, Oxfordshire OX9 2LP

A CIP catalogue record of this book is available from the British Library.

All rights in the photographs belong to the individual photographers who have made a non-exclusive assignment to Aves Press permitting their usage in this book and in any related print. The photographers can be reached through Aves Press.

All other rights are reserved. No part of this work may be reproduced or used in any form without written permission of the publishers and this applies to photographic, electronic or mechanical means, including photocopying, record, taping or information storage and retrieval systems.

Front cover: Wheel window, All Souls Church, Susans Road

eastbourne in detail

contributors

Concept	Richard Deane "Salisbury in Detail" Edward Dickinson	
Text	Richard Crook	
Graphic design	Nicholas Howell	
Photographers	John Crook Richard Crook Za Crook David Cross Dorothy Dickinson Nicholas Howell Rhys Hutchings, Sussex Downs College	Jeff Klepper Chris Leach David Skinner Barry Southon Janet Taylor Jonathan Webley
Typescript	Za Crook Dorothy Dickinson	Kim Read Janet Taylor
Steering Committee	Edward Dickinson, Chair Susan Body Richard Crook Dorothy Dickinson	 Nicholas Howell Chris Leach David Tutt

acknowledgements

Nigel Allchorn, Jack Brownell, Jacqueline Brunjes, Will Callaghan, Laura Cartledge, Anthony Crook, Clare Dales, Michelle Draycott, Christina Ewbank, Annemarie Field, Nat Gonella, Steven Goss-Turner, Ele Graham, Diana Guthrie, Roger Hilton, David Jones, Lionel Jones, Chris Jordan, Gill Kemp, Liza Laws, Rebecca Madell, Emma Morris, Joanne Naunton, Michael Partridge, Bill Plumridge, Dennis Scard, Tony Welton, Steve Woods, All Saints Chapel, Berkeley Homes, Friends of Manor Gardens and Gildredge Park, Sussex Downs College, University of Brighton. And for help with presentations at the Eastbourne Heritage Centre: Alan Davis, Diana Guthrie, Nicholas Howell, Paul Jordan, Chris Leach, Brian Peacock, Pauline and Graham Swift, Steve Woods.

contents

		page
Foreword by the Rt. Hon. The Earl of Burlington		1
Introduction by Richard Crook		2
01	Street furniture	6
02	Statues	16
03	Windows	22
04	Doors and porches	38
05	Door furniture	50
06	Brick and flintwork	58
07	Stonework	66
08	Ornate plasterwork	82
09	Decorative ironwork	94
10	Terracotta	106
11	Stained glass	114
12	Balconies, canopies and shelters	124
13	Decorative tiling and mosaics	134
14	Plaques	142
15	Clocks and sundials	152
16	Signage	158
17	Chimneys	164
18	Roof embellishments	172
19	Turrets and towers	186
20	Weathervanes	194
Afterword by Edward Dickinson		201

sponsorship

major sponsors

EDEAL: Eastbourne and District Enterprise Agency Ltd.

The Eastbourne Society

The Grand Hotel

Hawk Builders and Shopfitters Ltd.

John D. Clarke Architects

M.W. Pyle Roofing Ltd.

Richard Rager & Peter Roberts

sponsors

Affinity Architects Ltd.

Barwells Solicitors

Bede's School

Trevor Body

Brewers Decorator Centres

Building Design Architects Ltd.

Mr Cherry Picker Ltd.

Clarke Roofing Ltd.

Eastbourne Chamber of Commerce

Christina Ewbank

Gardners Books

Harveys Brewery

Hedley Visick Ltd.

HTP Structural Engineers

The Lansdowne Hotel

Lawson Lewis Blakers Solicitors

La Locanda Del Duca Ristorante Italiano

Mayo Wynne Baxter Solicitors

J. H. Payne & Sons Ltd.

G. E. Pearce Tree Services Ltd.

Gordon and Mary Piggott

Polegate Roofing Ltd.

Same Day Express

Stredder Pearce

Switchplane Ltd.

Tingley Commercial

Douglas Vernon

Visick Cars Ltd.

foreword

THIS SPLENDID BOOK has opened my eyes to the many often-overlooked details of the town, which was largely created in the Victorian age by my great great great great grandfather the 7th Duke of Devonshire. Before assuming that title, when he was the Rt. Hon. Earl of Burlington, he lived at Compton Place and from his enjoyment of living there came his very substantial involvement in the development of the town's plans and bye-laws. One of the really special legacies of his interest is the preservation of a frontage to the sea of properties, almost devoid of shops, which distinguishes the town from its neighbours along the coast.

"Never go East of the pier, dear" I was told once. However, the town has changed over the years and this book is testament to the joy to be had by heading east, west, in all directions in fact, throughout Eastbourne, to discover for yourself the visual delights that it holds. There are over 100 buildings in the area that are listed on account of their architectural merit, but Richard Crook and Nicholas Howell give us a wonderful reminder to dwell not only on the grander offerings but also the minutiae and smaller elements that defy their diminutive stature by adding significantly to the character of the place.

As a photographer myself I learnt early on to keep my eyes open and to look out for the unexpected. I am particularly pleased to see the quality of the illustrations and the treatment of all of them in colour, which, it was pointed out to me, might remind readers of the enviably high number of hours of sunshine per year that Eastbourne enjoys.

The Rt. Hon. The Earl of Burlington

introduction

EASTBOURNE has been described as an outstanding example of aristocratic seaside development. As Peter Brandon pointed out in his book *The Sussex Landscape,* "it was the 7th Duke of Devonshire's achievement to convert a straggling series of hamlets into a handsome, well laid-out town with miles of streets ornamented by noble terraces and princely mansions".

In the early 1800s, the parish was largely rural, with four separate centres of population. East-Bourn now known as Old Town, developed from very early times as an agricultural community. Its wealth in the medieval period is shown today by the splendid parish church of St Mary and the Lamb Inn. The other centres of Meads, South Bourne and Sea Houses were little more than hamlets. All of these have now been swallowed up in the modern resort of Eastbourne.

Eastbourne was late in its development as a seaside town. By 1830, the total population was only 3,000. During the 18th century, the town's only claim as a watering place of note had been in 1780 when Prince Edward (the father of Queen Victoria) stayed at a converted windmill known as the Round House, which once stood near the entrance to the Pier.

During the mid-19th century, the parish of Eastbourne was largely owned by two major landowners. One was the 2nd Earl of Burlington, William Cavendish, who became the 7th Duke of Devonshire in 1858, the other was Carew Davies Gilbert.

William Cavendish (1808 to 1891) was "one of the finest flowers of the Victorian nobility", being chairman of the Royal Commission on Scientific Instruction and the Advancement of Science, Chancellor of the Universities of London and Cambridge and President of the Royal Agricultural Society. He had inherited two extensive estates from his grandfather in 1834 when he became the 2nd Earl of Burlington: the Sussex estate at Compton Place in Eastbourne and the Lancashire estate at Furness Abbey with Holker Hall. In 1858, he became 7th Duke of Devonshire on the death of the 6th "bachelor" Duke, his second cousin, thus inheriting estates in Ireland (Lismore Castle), London (Chiswick and Devonshire Houses) and Derbyshire with Chatsworth House, thus becoming one of the wealthiest landowners in the country.

The second major landowner, Carew Davies Gilbert, was an agricultural innovator and one-time President of the Royal Society. He was also Lord of the Manor of Eastbourne. The main part of the Gilbert estate, about a quarter of the total land, was situated on the eastern side of the parish.

William Cavendish as the then 2nd Earl of Burlington had begun his development proposals at Eastbourne in 1838 by commissioning the famous architect Decimus Burton to draw up the first building plan on the Compton Estate for a new town to be called Burlington. Burton had worked with the 6th Duke and Joseph Paxton on a vast greenhouse called the Great Conservatory at Chatsworth House and also designed the Palm House at Kew Gardens. In London he had designed the Wellington Arch and the Hyde Park Screen together with many classical houses. However, at that time, the Earl seemed reluctant to spoil his rural seclusion especially after the death of his beloved wife Blanche in 1840 and his pressing involvement with his industrial developments at Barrow-in-Furness. The early plans at Eastbourne were not therefore carried out except for Burton's design for the Trinity Chapel, completed in 1839 (now much enlarged and called Holy Trinity Church).

The foundation stone of the new town of Eastbourne was laid in the spring of 1851, the first building being the Burlington Hotel followed shortly afterwards by Cavendish Place, Victoria Place and Cornfield Terrace. This early development could not have taken place without the arrival in Eastbourne of the railway which opened on 14th May 1849 when the London, Brighton and South Coast Railway extended a branch line from their main Brighton to Hastings line at Polegate (after much persuasion by the then Earl of Burlington who was a director of the Furness Railway in Lancashire).

When William Cavendish became the 7th Duke of Devonshire, he appointed Henry Currey as his architect who then produced the first comprehensive "Design for the Layout of the Duke of Devonshire's Estate for Building Purposes" in 1859. This plan was carried out with only minor alterations. Currey was born into a distinguished family and was the son of Benjamin Currey, agent and legal advisor to the Dukes of Devonshire. His visionary layout for Eastbourne was based on wide French boulevards and the Italianate (especially Venetian) architecture he had studied during his Grand Tour of Europe. The Eastbourne boulevards would be planted with elms, paved in local brick and softened by grass verges. The boundary walls would be local stone and flint. All this was paid for by the 7th Duke. The building plots were let out to local contractors who had to submit architectural plans to the local estate office for approval.

The local agent to the Duke of Devonshire was George Ambrose Wallis who dominated the late Victorian scene in the town. The provision of services (sewerage, gas and water) was vital for the development of a successful resort and in 1860, at the age of 20, George Wallis was taken on as 'resident engineer' to the Eastbourne Waterworks Company which was owned by the Duke of Devonshire. He subsequently became the Duke of Devonshire's local agent in 1864, running the building estate in Eastbourne. He also ran with his brother William the building firm of Wallis and Wallis that carried out all the major Ducal contracts including drainage and sewers, waterworks, the Western Parades and for a dramatic corniche called Upper Duke's Drive. Wallis was Chairman of the Local Board and became first Mayor of Eastbourne in 1883. Wallis also designed and built the popular Devonshire Baths whose main architectural feature was a Lombardo-Venetian tower which cleverly disguised the boiler chimney and water tanks. The Baths have sadly been demolished as have the grand mansions he built for himself at Holywell Mount and Fairfield Court.

In 1872, Henry Currey produced a second plan for the further development of the Duke of Devonshire's estate up to Meads which became the "Belgravia of Eastbourne". He designed the Western Parades and Grand Parade which *The Builder* magazine described as "one of the most picturesque in the country". Henry Currey was the architect of a number of local buildings including Eastbourne College, the Winter Garden, Queens Hotel, Devonshire Park Theatre, and a number of stucco terraces. The resort was carefully zoned with a formal town centre and seafront consisting of Italianate stucco buildings with low-pitched slate roofs. In Meads, a more Arts and Crafts style of architecture was adopted using local red brick with steeply-pitched tiled roofs, vertical tile hanging, turrets and gables. This gave rise to the epithet "many gabled Eastbourne".

During the late Victorian period, Eastbourne was the most rapidly-expanding resort in Sussex. The population doubled to 21,500 in the five years 1871 to 1876. One might have expected a chaotic result but, due to careful control of development, the result was outstanding and has been described by the late Christopher Hussey in *Country Life* as "a model of early planning which should be recognised as a masterpiece of its genre".

The sewerage system led to an outfall at Langney Point which was opened at a lavish event, with top hats worn and a procession of 60 carriages attended by the Duke and Wallis on 3rd May 1867. Standing in the front row, the 7th Duke of Devonshire takes centre stage and holds a ceremonial staff (could it be the sluice lever?). The Reverend Thomas Pitman, who was Vicar of Eastbourne from 1828 to 1890, stands by the Duke's left arm, and George Ambrose Wallis is second to the right of the Duke.

Eastbourne c 1876 showing in the foreground All Saints Convalescent Hospital (1869) and across the fields St John's Church and its vicarage (1870) before Meads was developed as the 'Belgravia of Eastbourne'. The view also depicts the Wish Tower (1804), the pinnacled Holy Trinity Church tower (1839), the Pier and St Saviour's Church spire (both completed 1872), and the Winter Garden which was completed in 1876.

This book seeks to celebrate Eastbourne's "detail", the small parts that subliminally add so much to our buildings and infrastructure and which are often overlooked. As with music, a composition is often more than the sum of its parts and it is often difficult to pinpoint why one enjoys a favourite tune or finds a building particularly attractive. In architecture, it is the balance of proportion, scale, rhythm, choice of building materials and the craftsmanship which assembles them and, in particular, carefully chosen ornamental detail which can lift the spirit and be an inspiration for future generations.

The Victorians in particular had an extravagant passion for detail. They felt that any building, however humble, should delight the eye and stimulate the senses. Eastbourne has a wealth of details to enjoy. Details that evoke the mood and fun of the seaside. Decorative cast ironwork abounds from filigree railings and ornate lamp posts down to the smallest foot scraper and coal hole. The classical town centre seafront buildings have elegant stucco façades with moulded string courses, rosettes, garlands, brackets and enriched details of every description. The interplay of light and shade from canopies and delicate timber fretwork is a delight on a bright summer's day. The later vernacular style buildings of Meads and Upperton have elaborate carved gables, bargeboards, finials and a variety of decorative roof ridge crestings. Enchanting porches, stained glass windows, brick patterned walls and pillared doors appear at every turn. The Victorians knew that detail subconsciously elevated the humblest building into a work of art and as John Ruskin noted "architecture proposes an effect on the human mind, not merely a service to the human frame".

The high standard of visual value set by the Victorians was adopted by subsequent architects and designers in the Edwardian period in particular. The 1920s and 1930s also saw some outstanding work and even in later periods, Eastbourne has not suffered so much of the blight that affected other towns and cities when the starker Modern Movement of architecture became established. In the 1960s, we have the outstanding Congress Theatre and more recently the award-winning Towner Gallery which continue to uphold the town's high design standards.

Much of the ornament that once graced Eastbourne's buildings is missing and this book seeks not only to open up visitors' and residents' eyes to the richness of the town's details but will hopefully inspire the owners of buildings to reinstate missing details. In addition, we hope it will act as a catalyst for future designers and architects to adopt the town's motto "Meliora Sequimur" which means "let us follow better things".

01

street furniture

Eastbourne displays a wealth of interesting details in its street furniture. Henry Currey laid out the streets in a grid pattern for the 7th Duke of Devonshire and in a similar vein, Nicholas Whitley laid out the Upperton Estate for Carew Davies Gilbert. No expense was spared in ensuring the detail was of the highest quality and a rich palette of materials was specified. Red paving bricks from the High Brooms brick fields in Tunbridge Wells were used for pavements and on the seafront parades so that walking the promenade was often referred to as "walking the bricks". The brick pavements were edged in Purbeck stone kerbs from Dorset and the highest quality streets all had grass verges.

All the streets were planted with Wheatley elms to create tree-lined boulevards reminiscent of Paris. The town soon became known as "Leafy Eastbourne". At junctions, granite setts were laid so that Victorian ladies' crinolines would not pick up the dust from the streets. The paving bricks were soft so hard Staffordshire Blue crossovers were used to delineate entrance driveways to the villas. Originally, the roads were untarmacked and in dry weather became dusty. The streets were sprayed with water from horse-drawn water carts which were filled from hydrants at convenient locations. There are still a few examples of hydrants on our streets.

01

The water hydrants were served by the Eastbourne Waterworks Company and there are a few surviving meter covers around the town. Under the roads, cast iron sewers were installed to the designs of the 7th Duke of Devonshire's resident engineer, George Ambrose Wallis. The drains were served by ornamental ventilation (otherwise known as "stink") pipes cast by Ham Baker Ltd. of Westminster. The sewerage system led to an outfall at Langney Point which was opened at a lavish ceremony attended by the Duke and Wallis on 3rd May 1867 (shown in the historic photograph on page 4). In Hampden Park there is one surviving street urinal cast in Scotland by the famous firm of Walter Macfarlane & Co. Inside is an inscription "please adjust your dress before leaving" (shown on page 160).

page 7: St John's Road, looking west

this page: "Stink" pipe base, corner of Latimer and Cambridge Roads

facing page:
1. Gentlemen's urinal, detail of exterior, Hampden Park
2. "Stink" pipe terminal, Silverdale Road
3. Water hydrant, Grassington Road

1
2
ENGINEERS
3

The early cast iron street furniture was cast locally by Eastbourne's Victorian ironmaster Ebenezer Morris who had a foundry in Lewes. His work can be seen all over the town in railings, gutter grids, kerb crossovers and in particular in the square-based lamp posts. The gutter grids are often fitted with a specially designed cast iron overflow weir (a unique Eastbourne feature to get around the problem that there were so many trees that the grids became clogged with leaves). Children hated these because their cricket and tennis balls disappeared down the drains!

this page: Morris gutter grid and overflow weir, Silverdale Road

facing page:
1. Morris kerb crossover channel, Pashley Road
2. Granite sett crossover, Grove Road
3. Eastbourne Waterworks Company meter cover, Marine Parade
4. High Brooms Brick Co. pavement brick, Trinity Trees
5. Paving/Staffordshire Blue crossover, Meads Road

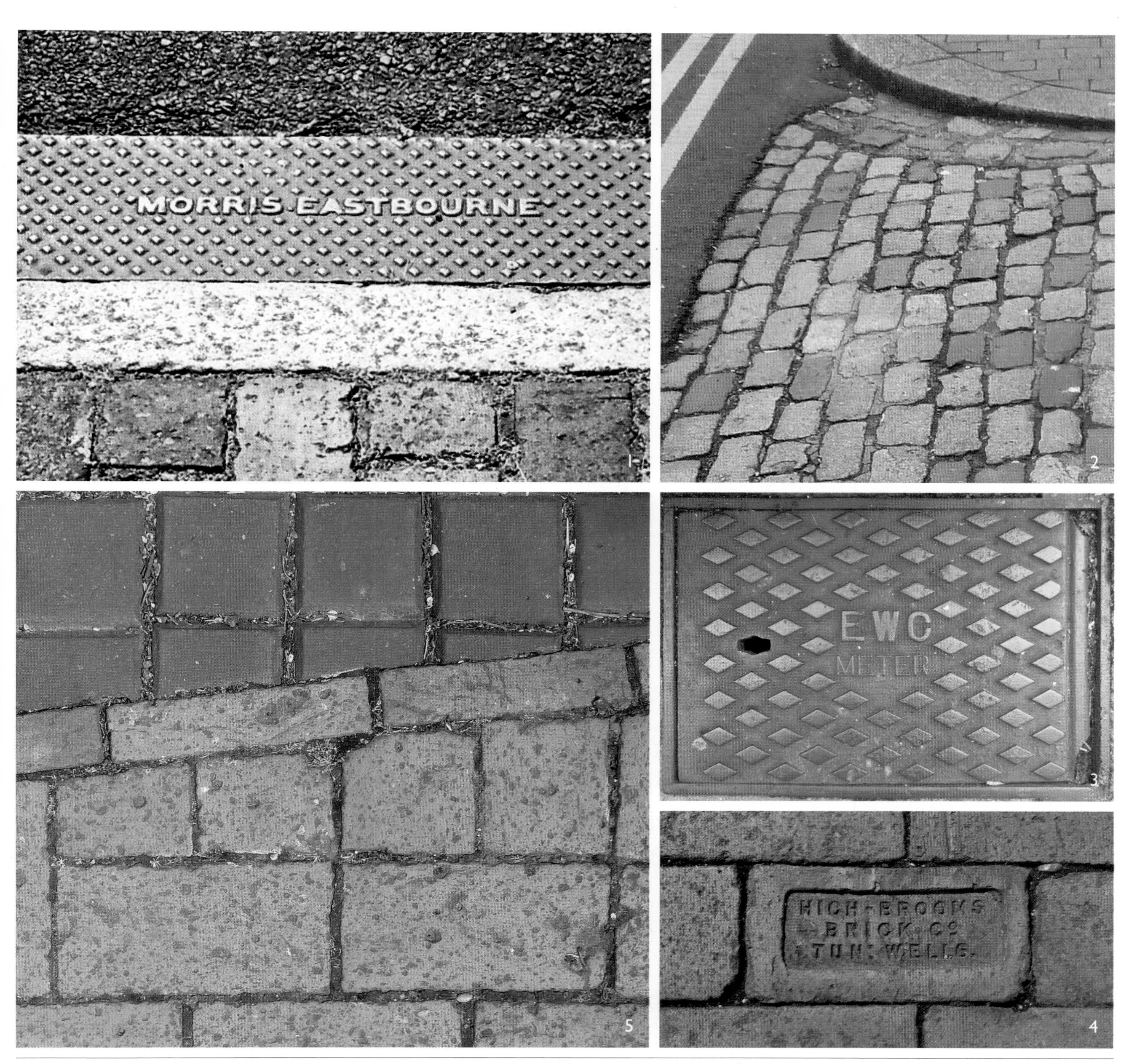
MORRIS EASTBOURNE
EWC
METER
HIGH BROOMS
BRICK Co
TUN. WELLS
1
2
3
4
5

MORRIS EASTBOURNE
1
2

A few of the square-based lamp posts (dating from 1852 to 1899) still survive. These were specially cast with a square base to house a large gas meter. The posts are beautifully ornamented with acanthus leaves, Greek anthemion, reed, and ball decoration. The original square gas lanterns were subsequently adapted for electric light from about 1900 and then replaced in the 1930s with electric swan-neck tops. The round-based ornamental lamp posts by John Every (who was apprenticed to Morris) are of a later date. The lantern on the original Morris lamp post (shown above and on the facing page) has ornament replicating detail found on a rare Eastbourne lantern in the Eastbourne Heritage Centre Collection.

this page:
1. Morris lamp post, Old Wish Road
2. Lantern detail, Christ Church, Seaside
3. Morris bollard, Carpet Gardens, Grand Parade
4. Every lamp post base, Watts Lane

facing page:
1. Morris railing, South Cliff
2. Morris lamp post, Christ Church, Seaside

Eastbourne has one of the best accumulations of Victorian post boxes in the country. There are eighteen VR (Victoria Regina) pillar boxes which date from 1887 to 1901 cast by A. Handyside & Co. of Derby, one VR wall box by W.T. Allen & Co. of London, and six anonymous boxes which have no royal cypher dating from 1883 to 1887. The town also has a number of interesting licence plates from where transport services were available, including: LPS for Luggage Porter Stand, HCS for Hackney Carriage Stand, BCS for Bath Chair Stand and GCS for Goat Chaise Stand. From a slightly later period, there is MCS for a Motor Charabanc Stand which can be found in King Edward's Parade by the Wish Tower boundary wall.

this page: Licence plates:
1. Luggage Porter Stand, Eastbourne College boundary wall, Grange Road
2. Hackney Carriage Stand, near Wish Tower, King Edward's Parade
3. Bath Chair Stand, near Wish Tower, King Edward's Parade
4. Goat Chaise Stand, near Wish Tower, King Edward's Parade

facing page:
1. Seafront bench near Redoubt Fortress, Royal Parade
2. Victoria Regina pillar box, South Cliff

02

statues

Statues decorate and enliven a town providing lasting memorials to heads of state, local heroes and civic dignitaries, and also sombre reminders of past wars. One of the oldest statues in Eastbourne is Neptune who reclines resplendently in the middle of the pond at Motcombe Gardens in Old Town. Near here rises the spring of the Bourne Stream from which Eastbourne gets its name. The pond originally covered the whole of the low lying ground and became the town's first water supply reservoir in 1844. The lead figure of Neptune was brought here from the large mill pond further down the stream in The Goffs. It is surprising to find in Eastbourne that there is no statue to Queen Victoria considering the expansion of the town in the period that bears her name. This can probably be explained by the fact that in 1901 (the year of Queen Victoria's death) an elaborate and expensive statue had been erected to the 7th Duke of Devonshire (who had died in 1891) "erected by the voluntary subscriptions of inhabitants of Eastbourne in commemoration of his generous interests in welfare and progress". It was designed by Sir William Goscombe John A.R.A. and unveiled by the Marquess of Abergavenny. The Duke is wearing robes as Chancellor of Cambridge University and is holding a book with the Eastbourne coat of arms on the cover (possibly this contains his development plans?).

02

The statue is the focal point at the seaward end of Devonshire Place, the central 80ft wide tree-lined boulevard of the town he created. The statue of his son, Spencer Compton, the 8th Duke of Devonshire, is by Alfred Drury and was erected on the Western Lawns in 1910. The 8th Duke was Mayor of Eastbourne from 1897 to 1898 and the statue portrays him, like his father, in robes as Chancellor of Cambridge University. He remembered fondly coming down to Compton Place with his father and wrote "as a child, I used to walk from Compton Place to the Wish Tower by a footpath by fields of waving corn, through that expanse of land which is now occupied by well-ordered streets and prosperous residences".

page 17: Neptune, Motcombe Gardens, Motcombe Lane

this page: 7th Duke of Devonshire, Devonshire Place

facing page: 8th Duke of Devonshire, Western Lawns

A striking memorial to the 2nd Battalion of the Royal Sussex Regiment by Sir William Goscombe John A.R.A. stands as a focal point at the end of Cavendish Place where it meets Grand Parade. Another prominent statue is the War Memorial which stands at the junction of Cornfield Road and Devonshire Place. This was designed by Henry C. Fehr of South Kensington and unveiled by General Lord Horne on 10th November 1920. The 6ft high bronze represents the Angel of Peace and Victory.

this page:
1. An unknown medieval knight, Pevensey Road
2. 2nd Battalion of the Royal Sussex Regiment Memorial, Grand Parade
3. St Michael slaying Satan as a dragon, Esperance Hospital, Hartington Place

facing page: War Memorial, Devonshire Place

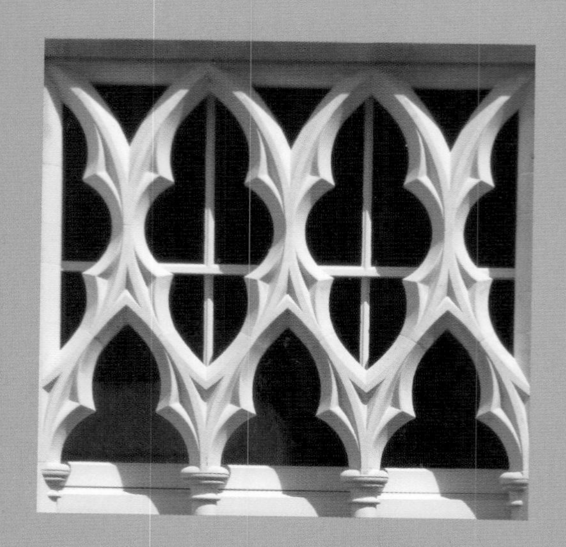

03

windows

Windows are the "eyes" of a building and the word is derived from the old Norse "vindr-uga" meaning "wind eye". As with eyes on a human face, the character of a building is very much dependent on its windows. Some of the earliest windows in the town are to be found in the medieval parish church of St Mary where, in the South Aisle, are good examples of early Decorated work with a design identical to that used in the choir screen in Canterbury Cathedral which was erected in 1304-5. Medieval pointed arch windows were copied and reinvented by the Victorians who developed a passion for Gothic architecture, especially in church buildings. These windows display biblical scenes in colourful stained glass framed in an infinite variety of delicate stone tracery with cusps, trefoils and quatrefoils. In Borough Lane, a house called Pilgrims has a medieval cellar and 16th century timberwork. The bay window there has hand-made geometric glass quarries set in lead cames and supported by saddlebars. Leaded lights were used in early windows as it was impossible at that time to produce large pieces of glass. Other examples can be found in the Lamb Inn and in the 16th century Old Parsonage behind St Mary's Church. From the mid-18th century right through to the end of the Edwardian period, the window of choice was the double-hung vertical timber sliding sash. This clever invention allowed flexible ventilation as each sash was carefully balanced by weights hidden in sash boxes at the sides of the windows.

03

page 23: St Mary's Church, Church Street

this page:
1. Old Parsonage, Ocklynge Road
2. Gildredge Manor, Borough Lane

facing page:
1. All Saints Convalescent Hospital, Darley Road
2. Pilgrims, Borough Lane
3. St Michael and All Angels Church, Willingdon Road

One of the earliest examples of the sliding sash in Eastbourne is at Gildredge Manor House. This was built around 1776 for the Reverend Dr Henry Lushington (Vicar of Eastbourne from 1734 until his death in 1779). The architect is unknown but could well be Sir Robert Taylor who was designing country houses in a very similar style at this time. The window shown is a beautiful example of the Georgian double-hung sash, set off by its red brick surround with a fine gauged brick arch head. The slender glazing bars are finely moulded and the perfect proportions are based on the great Renaissance Italian architect Andrea Palladio.

The Victorians adopted the timber sash window but industrial development in glass manufacture meant that they were no longer restricted to the small panes of hand-blown glass that had to be used by the Georgians. In Eastbourne, the windows of the Italianate classical buildings in the town centre and seafront are framed in an almost unbelievable variety of decoration with moulded architraves, sills and hood moulds. The decoration is however restrained and refined and less fussy than in other Victorian towns. A building in Devonshire Place has pairs of windows cleverly linked together by their architraves (shown above). Devonshire Place is the central axis of Henry Currey's 1859 plan and the imposing Italianate villas housed the cream of Eastbourne society and the window surrounds were suitably embellished to reflect this status with carefully proportioned pediments, brackets, husk garlands, bay leaf friezes and classical motifs.

this page: Sherwood Court, Devonshire Place

facing page:
1. Devonshire Place
2. Grand Hotel, King Edward's Parade
3. Spencer Road

1
2
3

The keystone of a window in Bolton Road is vermiculated like a worm cast and supported on moulded brackets (shown above). In the later residential areas of the town, a more vernacular Arts and Crafts style broke away from the tradition of classical architecture to find its own form of embellishment. One of the most profound changes in the town recently has been the gradual replacement of original windows with modern materials. Plastic cannot hope to recreate the subtlety of the original designs with their delicate mouldings and details. In addition, modern glass is featureless whereas the old glass has imperfections and figuring which subtly enlivens the "eyes" of the building.

this page:
1. Arlington Road
2. Furness Road

facing page:
1. Bolton House, Bolton Road
2. Blackwater Road

The windows of Eastbourne's public Victorian buildings are carefully detailed to suit their purpose. The Gothic arched windows of the Leaf Hall with their polychromatic brickwork are supported on Bath stone embellished capitals. This is the work of local architect, R. K. Blessley, who also designed the Grand Hotel in a very different classical style. The banks adopted a classical style while Eastbourne College looked further back to the Tudor period for its inspiration with leaded light windows. At All Saints Convalescent Hospital in Meads the architect Henry Woodyer employed the spirit of medieval English Gothic architecture but he did not copy and his designs are utterly unique. He had a passion and an eye for small detail and the windows with their flamboyant reticulated tracery resemble those found in an early Venetian Palazzo.

this page: Leaf Hall, Seaside

facing page: All Saints Convalescent Hospital, Darley Road

page 32: South Street

page 33:
1. Former bank building, corner of Gildredge and Terminus Roads
2. Eastbourne College, College Road
3. Peregrine House, Compton Place Road

1

3

2

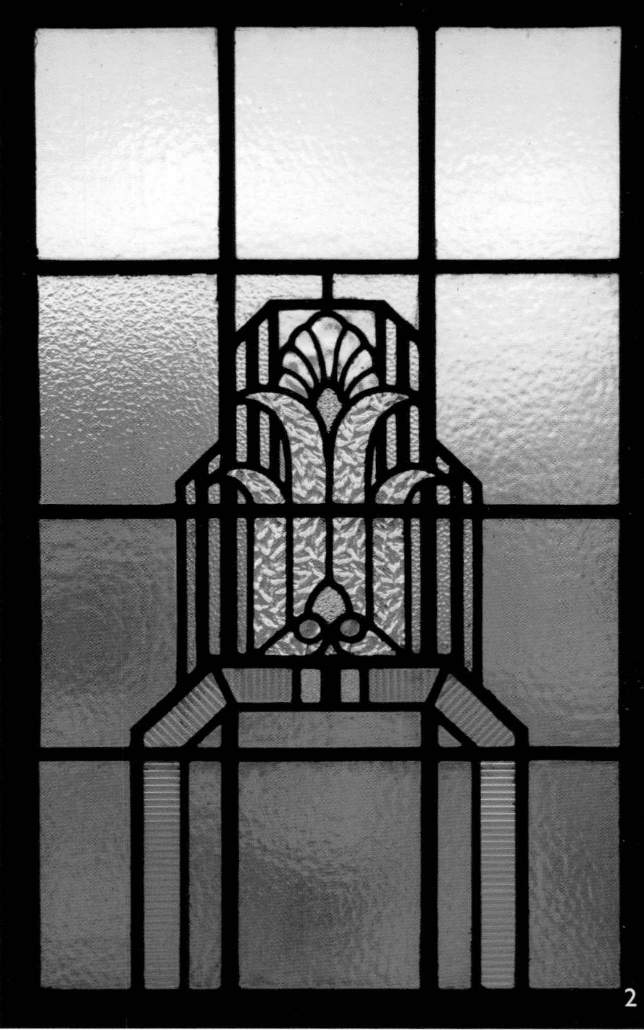

In the 1920s, windows developed to express the spirit of that age. The stream-form shapes surrounding the window of the former Police Station in Grove Road and the Art Deco glazing found in Pearl Court and at the former Tally Ho public house in Church Street are very much period pieces. More recently at the Birley Centre by Miller Bourne Architects and in the award-winning Towner Art Gallery by Rick Mather, we have striking examples of the sculptural forms of fenestration employed by modern architects.

this page:
1. Former Tally Ho public house, Church Street
2. Pearl Court, Devonshire Place

facing page: Former Police Station, Grove Road

this page: Birley Centre, Carlisle Road

facing page: Towner Art Gallery, Devonshire Park, College Road

04

doors and porches

The entrance door to a building has been throughout history a prime focal point in architecture. In Eastbourne, we have some beautiful examples on display such as the fine entrance at Gildredge Manor where the Tuscan porch with its dentil decoration is supported on Doric columns and pilasters. One of the earliest doorways in the town is the early 14th century inner door to the South Porch at St Mary's Church, Old Town. The Reverend W. Budgen in his book *Old Eastbourne* suggests that the heads supporting the hood mould represent Baron Bartholomew de Badlesmere and Margaret his wife who were granted the Manor of Eastbourne in 1308 by King Edward II.

The desire of the Victorians to go back to the medieval period for church architecture is represented in the 1859 Gothic arched entrance door at Christ Church in Seaside by Benjamin Ferrey (a pupil and biographer of the famous Augustus Welby Pugin) where the hood mould is supported by two majestic angels.

04

page 39: Gildredge Manor, Borough Lane

this page:
1. South Porch inner door, St Mary's Church, Church Street
2. Christ Church, Seaside

facing page: All Saints Convalescent Hospital, Darley Road

Henry Woodyer (a pupil of the famous architect, William Butterfield) at All Saints Convalescent Hospital once again uses his highly-inventive genius in the entrance porch. The hospital is constructed of local Sussex brick with Bath stone dressings and the design is highly expressive. The hospital was built for the All Saints Sisters of the Poor, an Anglo-Catholic community founded in 1851 by Harriet Brownlow Byron to help London's poor and sick recuperate by the seaside. This religious community grew out of the Oxford Movement founded in the 1830s which sought the revival of monasticism within the Church of England. Its aims were piety, chastity, obedience and justice and included the revival of active charity work.

No's 38 - 41
& 43 - 47

this page:
1. Greystone House, Meads Road
2. Hartington Place

facing page:
1. Chatsworth House, Devonshire Place
2. South Street

There are a few examples in Eastbourne of fine Georgian buildings with delicately detailed entrance door surrounds such as Greystone House in Meads Road which has a six-panelled door with fanlight over surmounted with a semi-circular pediment supported on double capital pilasters. As might be expected, the Victorians lavished attention on their entrance doors and porches. In Eastbourne's classical stucco terraces and villas of the town centre, the main entrances were almost always positioned above street level (and therefore above the hoi polloi) up a formal flight of steps leading to porticos of different styles as shown in the photographs. Some had Doric columns, others highly-decorated Corinthian columns. Others have porches of cast ironwork. These elegant buildings were created for the wealthy middle classes and were carefully planned to the social requirements of the day. They were run by a veritable army of cooks, nannies, butlers and maids. To run an acceptably "genteel" and "proper" household, three servants were the absolute minimum. The home was central to the middle class view of life: a place where standards of behaviour and taste were nurtured, where ugliness and vulgarity were excluded. Devonshire Place was quite the finest road in Eastbourne and at Chatsworth House for example, the imposing entrance led to formal reception rooms while in the basement cook and scullery maid would be slaving over a hot range.

The later Meads mansions of Currey's 1872 plan were set in extensive grounds and were built for the upper middle classes who wished to emulate, in a small way, the aristocratic lifestyle of Compton Place and its estate. They are of a picturesque vernacular style built of local red brick, flint and Greensand embellished with Bath stone dressings. Some houses had elaborate timber porches with geometric glazing and others had turned timber pilasters. The buildings present a delightful contrast to the more formal villas of the town centre. The Victorian Society considers Meads to be a remarkably intact and outstanding example of a late 19th century planned resort.

this page:
1. Sherwood Court, Devonshire Place
2. Saffrons Road

facing page: Staveley Court, Staveley Road

STAVELEY COURT

The high standards set by the Victorians continued in the Edwardian period and in the 1920s and 1930s. There are exceptional examples of entrance doors such as at Pearl Court in Devonshire Place built for the Pearl Assurance Company in 1936 by F. C. Benz. It is the town's finest Art Deco building which might come straight out of an Agatha Christie Hercule Poirot novel. One can imagine during this period the residents being picked up by "Silver Wings", a high class local taxi service of highly polished Rolls-Royces driven by uniformed chauffeurs.

this page:
1. Eastbourne Railway Station concourse
2. Saffrons Road

facing page: St John's Road

this page:
1. Carew Road
2. Granville Road

facing page: Pearl Court, Devonshire Place

PEARL COURT
FLATS 9-16

05

door furniture

The embellishment of the entrance door was of fundamental importance. In the early Victorian period, there was no electricity and front doors were furnished with a bell pull connected by wires and levers to the main entrance doorbell which rang noisily from a swivelling spring. A good example of this is the front door bell at Eastbourne Heritage Centre (shown on page 54). Internally, principal rooms were all furnished with bell pulls connected to bells and swinging flags to summon the chief maid of the house to the correct room. The alternative to the bell pull at the front door adopted in the Georgian and Victorian periods was the door knocker. Later on, the electric bell push was adopted and embellished in suitable stylistic detail of the day.

The doors as St Saviour's Church (above) show the attention to detail lavished on door furniture on church doors. George Edmund Street and Henry Woodyer, the architects of St Saviour's and All Saints Convalescent Hospital (shown on page 52), designed not only the buildings but also the ironwork where each door handle and hinge is of a different design. Front door letterplates in particular give the opportunity for house owners to add their own highly personal stamp to their property.

05

page 51: St Saviour's Church, South Street

this page:
1 & 2. All Saints Convalescent Hospital Chapel, Darley Road

facing page:
1. St Saviour's Church, South Street
2. St Michael and All Angels Church, Willingdon Road

1

2

this page:
1. Eastbourne Heritage Centre, Carlisle Road
2. Meads Street
3. Vicarage Road

facing page:
1. Carlisle Road
2. Eastbourne Railway Station concourse
3. The Winter Garden, Compton Street
4. Carlisle Road

1

4

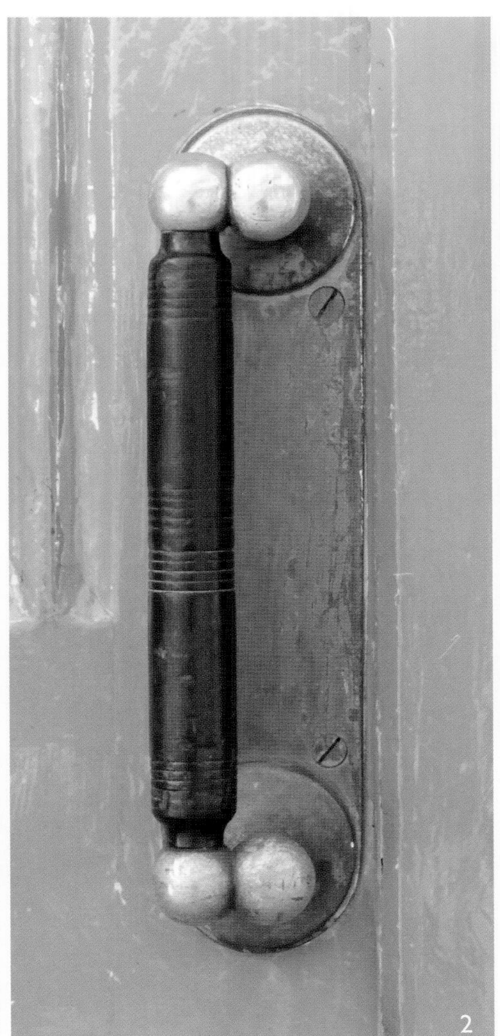

2

3

page 56:
1. Compton Street
2. Avenue Mansions, Elms Avenue
3. Borough Lane
4. Greystone House, Meads Road
5. Grove Road Chambers, Grove Road

page 57:
1. Crown Street
2. Manor Hall, Borough Lane
3. Saffrons Road

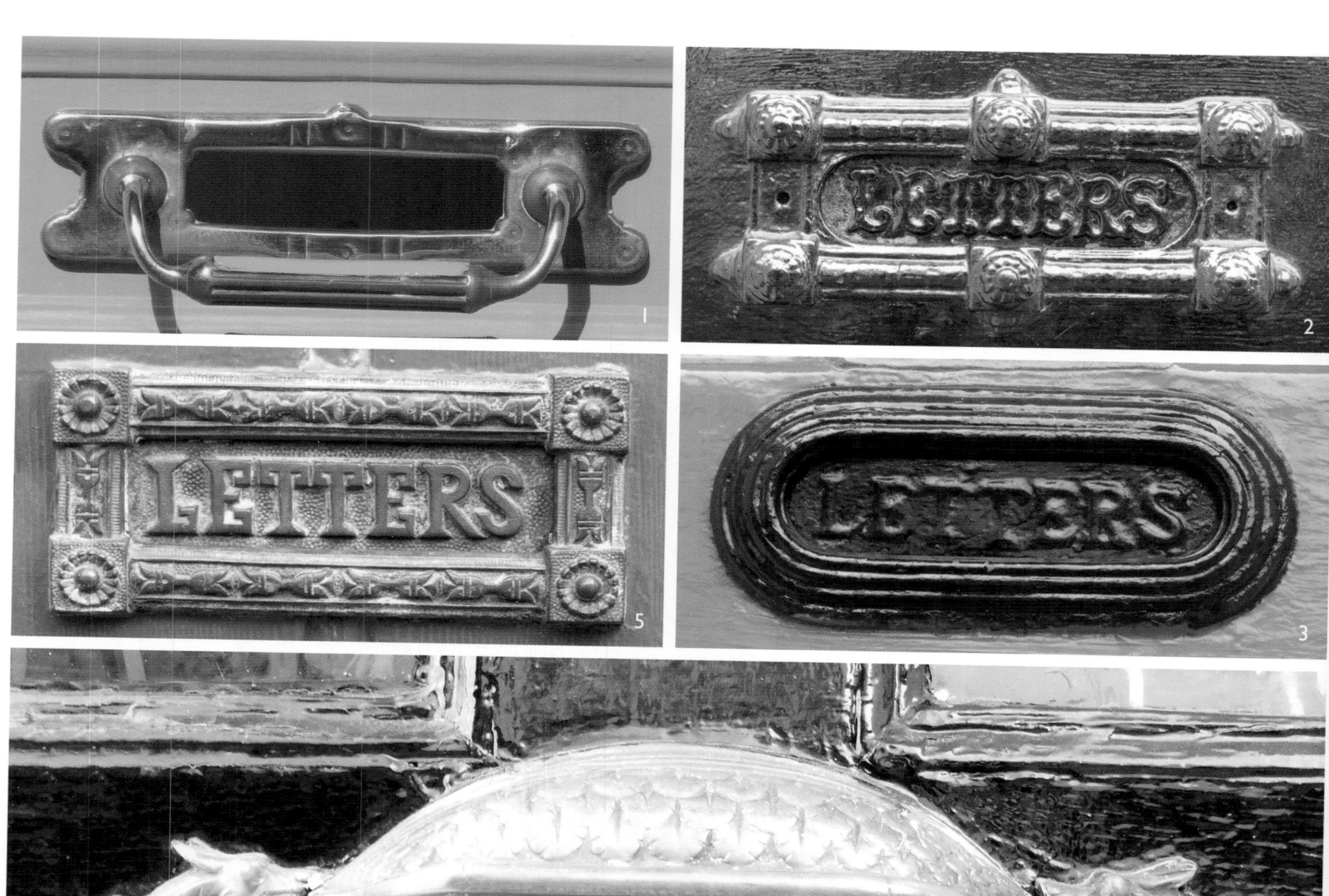
1
LETTERS
2
LETTERS
5
LETTERS
3

4

1

2

3

06

brick and flintwork

Bricks and tiles were first made in Sussex nearly 2,000 years ago. In Eastbourne, the first mention in the directories of a brickmaker was Samuel Gravett in 1796. Bricks were made for the Martello Towers locally and in the Victorian period brick fields were developed in Willingdon, Seaside, Roselands and on the site of present day Firle Road and Bourne Street to be gradually replaced by housing. The Eastbourne Brick Company Limited had a registered office in Seaside Road. The bricks were embossed with an EASTBOURNE stamp in the "frog" indent (shown on page 60). Surprisingly, James Peerless, the most notable builder in Victorian Eastbourne (e.g. the Town Hall, All Souls Church and St Saviour's Church) had his own brickworks in the town. The bricks produced used "brick earth" from the local sand and clay, hand-thrown into moulds and baked in brick kilns. The Eastbourne bricks are deep red or plum with slight colour and texture variations that make them so bright and cheerful, especially on a sunny day. This distinctive colour adds so much to the character of Eastbourne's public buildings, churches and vernacular Meads mansions. The bricks were laid in lime mortar with thinner joints than now used and laid in various bonds. English Bond and Flemish Bond were the most common where the relationship between stretchers (i.e. whole bricks) and headers (i.e. bricks laid end on) was carefully defined. Special bricks at the corners were needed to keep the bond known as King and Queen Closers.

06

1
EAST-
BOURNE
4
2
3

The bricks were soft and could be precisely shaped and rubbed to produce gauge brick arches with very tight joints (approximately 2mm) bedded in pure white lime mortar. Most mortar used lime up until the 1920's when the use of cement gradually superseded it. Some bricks were specially patterned using shaped moulds, other special bricks known as "rubbers" were made using finer and purer clay which could be sawn and sanded into the required shape by hand including actually carving the bricks on site. Buff, black and grey bricks were used from other brick fields in the country and were sometimes used together to form polychromatic brickwork such as used in the Leaf Hall (J. K. Blessley, architect).

page 59: Greensand, flint and brickwork, Blackwater Road

this page: Polychromatic bricks, Leaf Hall, Seaside

facing page:
1. Moulded brick, Arlington Road
2. Tuck pointing, College Road
3. Hand carved bricks, Arlington Road
4. Eastbourne brick

Polychromatic brickwork was also used in the magnificent Bedfordwell Pumping Station designed by Henry Currey for the Eastbourne Waterworks Company. The foundation stone is shown on page 144. The building was opened by the Prince and Princess of Wales during their royal visit in 1883. The finest brickwork such as that used in the villas of College Road and Eastbourne Heritage Centre (built by Wallis and Wallis in 1880 as the manager's house for the Devonshire Park and Baths Company) had their joints coloured to match the bricks and were then penny-pointed by means of a metal disc which was then infilled with white lime tuck pointing, a laborious but finely-detailed method of finishing the joints.

this page: Bedfordwell Pumping Station, Bedfordwell Road

facing page: South Street

The local flints from the sea (boulder flints) and from the fields of the South Downs (field flints) were employed in a variety of ways. Some were laid randomly, others in courses using carefully selected similar sized flints. This variety can be found at Holy Trinity Church, designed by Decimus Burton and completed in 1839 (shown on the facing page). Some flints were "knapped" (i.e. broken in half to expose the black core) and the finest works such as found on the gateposts of Compton Place and in the Paradise Belvedere (shown on page 130) were "knapped and squared" where the flints were broken in half and then carefully squared in shape by a skilled flint knapper.

this page: All Saints Convalescent Hospital, King Edward's Parade

facing page: Holy Trinity Church, Trinity Trees

07

stonework

As we have already seen, local brick and flint was used in the construction of buildings in the town. Eastbourne Greensand, which comes in various shades of green, has also been used for centuries in the area. Eastbourne Greensand is the local name given to Upper Greensand which underlies the chalk and notably forms the reefs at Holywell. This stone was used in the construction of Pevensey Castle and it is thought to have come from a quarry where the Cavendish Hotel now stands. At St Mary's Church in Old Town, the tower is entirely built of Eastbourne Greensand (shown on page 153) as are many of the walls, where it is combined with flints and also used in the stone dressings of window tracery, doors, string courses and quoins (shown on page 23).

In the Victorian period, the Greensand was mainly used for boundary walls and for feature "panels" in buildings, often with brick surrounds and flint (shown on page 59). The local chalk (known as "clunch") was not normally used externally in buildings but sometimes employed internally, particularly in churches. As transport improved, stone was increasingly brought in from elsewhere especially from the Weald which has a beautiful buff coloured sandstone. It is interesting to see however that even in the medieval period, stone was brought from far afield including the finely carved cream Caen stone from France which is found in the pillars at St Mary's Church.

07

The town's most important examples of Victorian architecture, All Saints Convalescent Hospital, St Saviour's Church, and All Souls Church, all used Bath stone which can be finely carved. The carved stonework of All Saints Convalescent Hospital's Chapel of 1874 is outstanding especially in the chancel at the east end where the trefoil and dagger traceried windows burst through the roof at eaves level surrounded with a carved trellis grid of quatrefoils. The architect Henry Woodyer remains an obscure figure, avoiding publicity and joining no professional body. He was a man of means and spent most of the summer in the Mediterranean on his yacht, *The Queen Mab*.

The Early English style capitals in the entrance porch at St Saviour's display the highest quality of undercut carving (you can put your fingers in and behind the "foliage"). The shafts supporting some of the capitals are of Sussex "marble" from Petworth and the Weald. This splendid Tractarian church by George Edmund Street (architect of the Law Courts in the Strand) was opened in 1867 with the spire added in 1872.

page 67: St Mary's Church, Church Street

this page: St Saviour's Church, South Street

facing page: All Saints Convalescent Hospital Chapel, Darley Road

pages 70/71: Eastbourne Town Hall, Grove Road

1
2
3
4

Bath stone was also employed in the construction of the railway station which opened in 1886 by the London, Brighton and South Coast Railway where in the station concourse, one section of the foliated string course is carved as a fierce wyvern (a winged creature with dragon's head, wings, reptilian body, two front legs and a barbed tail!). This was adopted by the London, Brighton and South Coast Railway to support their coat of arms. Here, each keystone is of a different carved design. Eastbourne Town Hall used Portland stone from Dorset to complement its fine Eastbourne red brickwork. This building was designed by Birmingham architect W. Tadman Foulkes after an architectural competition and was opened in 1886. With its asymmetrical façade and 130ft clock tower, it is the most flamboyant and obviously Victorian building in the town.

this page: Eastbourne Town Hall, Grove Road

facing page:
1-3. Keystones, Eastbourne Railway Station concourse
4. Wyvern, Eastbourne Railway Station concourse

The leering figure (shown above) is one of many characters that decorate Folkington Manor which was designed John Donthorne in 1843. Next to it is a detail of a fine carving at Holy Trinity Church. Completed in 1839, it was built as a Chapel of Ease to the Parish Church by the famous architect Decimus Burton who had drawn up the initial development plans for Eastbourne for the then Earl of Burlington. The carved cartouche at Caffyns Motor showroom (1910), on the corner of Saffrons Road and Meads Road, was carved in Bath stone by local sculptor Charles Godfrey Garrard (shown in photograph 2 on page 79). Moving on to the 20th century, Portland stone was used again for the stylish Art Deco detailing of Pearl Court (1936) and in the sculpture by Davis on the side of Eastbourne Public Library (1964), (shown on pages 80/81).

this page: St Michael and All Angels Church, Willingdon Road

facing page:
1. Carved figure, Folkington Manor, Folkington
2. Carved acorn detail, Holy Trinity Church, Trinity Trees

this page:
1. Lifeboat Museum, King Edward's Parade
2. Blackwater Road
3. Grassington Road

facing page:
1 & 2. Grassington Road

page 78:
1. NatWest Bank, Terminus Road
2. Old Orchard Road

page 79:
1. Eastbourne Town Hall, Grove Road
2. Corner of Saffrons and Meads Roads
3. Former Police Station, Grove Road

page 80:
1 & 3. Pearl Court, Devonshire Place
2. Burlington Court, Burlington Place

page 81: Eastbourne Public Library, Grove Road

1

2

1

2

3

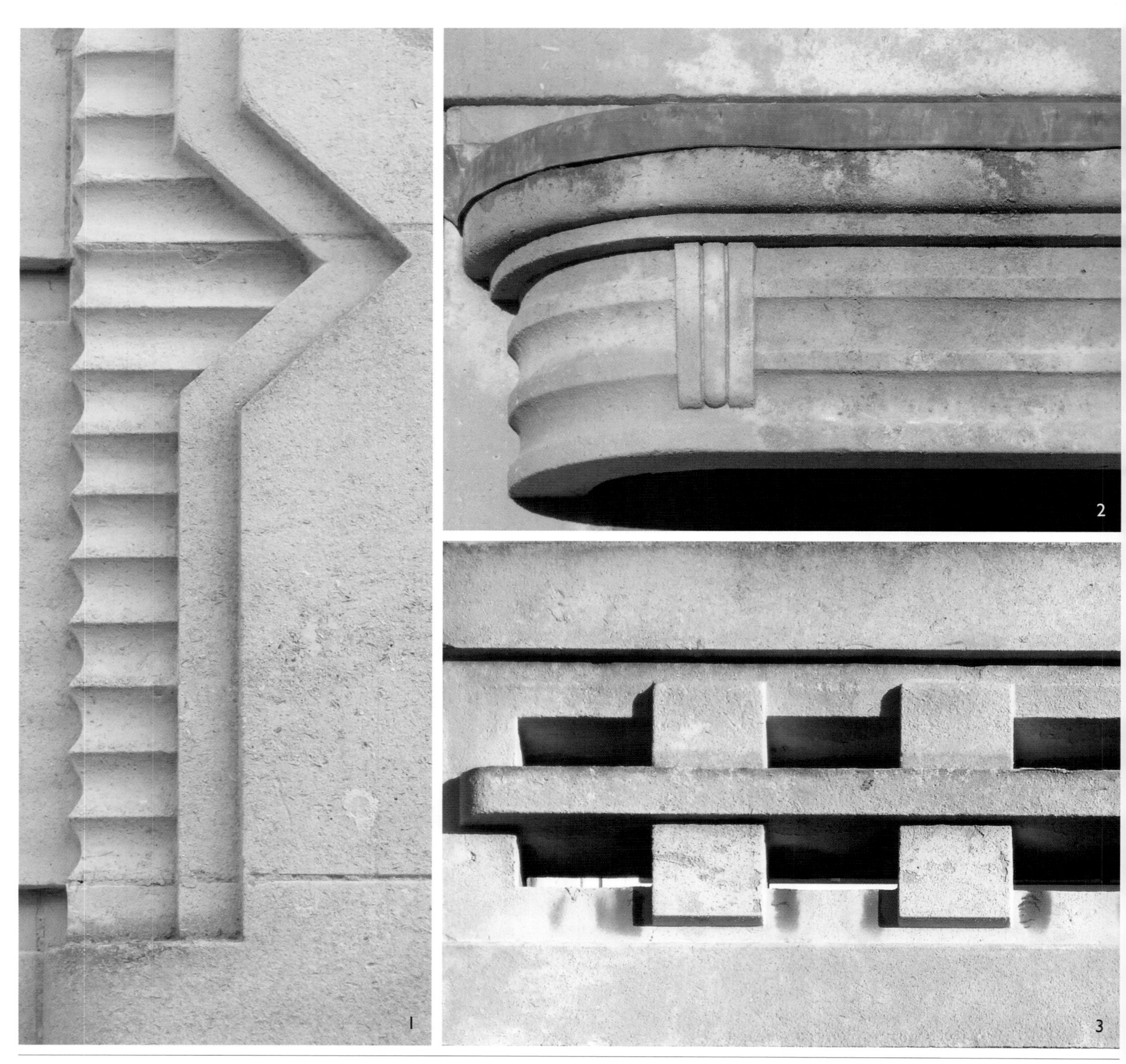
1
2
3

08

ornate plasterwork

Careful development control by the agents of William Cavendish, 7th Duke of Devonshire, ensured that the terraces, villas and hotels of the town centre would be in a formal, Italianate classical style. These stucco or plastered buildings display a wealth of carefully and correctly detailed ornament executed to perfection by incredibly skilled craftsmen. Some of the delicate mouldings including ornamental brackets, capital tops and other decoration were pre-cast before being brought on site. Other linear ornaments such as cornices, string courses and window surrounds, were run on site using zinc templates run on timber battens to create the correct profiles in the wet plaster. Whilst this style of architecture represents a cohesive whole, every building displays different ornamentation to achieve an immensely rich and varied townscape. For example, the main façade of the Cavendish Hotel by T. E. Knightley (architect of the Queen's Hall, London) employs Egyptian motifs with anthemion triglyphs, egg-and-dart mouldings, lotus flower decorations as well as incised decoration known as pargetting. All this is surmounted by a Sphinx's head. The hotel was commenced in 1866 but due to the difficulty in getting tenants was not finished until 1873.

08

page 82: Burlington Hotel, Grand Parade

pages 83, 84/85: Cavendish Hotel, Grand Parade

this and facing page: Grand Hotel, King Edward's Parade

There are three main "orders" of architecture which are easily recognised by looking at the column tops, known as "capitals". Doric with a plainly moulded capital, Ionic with curly voluted capital like rams horns as found on the Burlington Hotel and the highly-ornamental Corinthian capital with acanthus leaf and central fleuron (flower head) decoration which was invented in Corinth. An even more decorative capital called "Composite" combined Ionic and Corinthian and was used at the Grand Hotel, designed by R. K. Blessley and opened in 1876. The Grand has that exclusive air typical of Victorian Eastbourne and with its brilliant façades and precise enrichments (cornices, brackets, urns and columns), it gives the appearance of a veritable palace built of icing sugar. It is here that Debussy orchestrated *La Mer* during his stay in 1905.

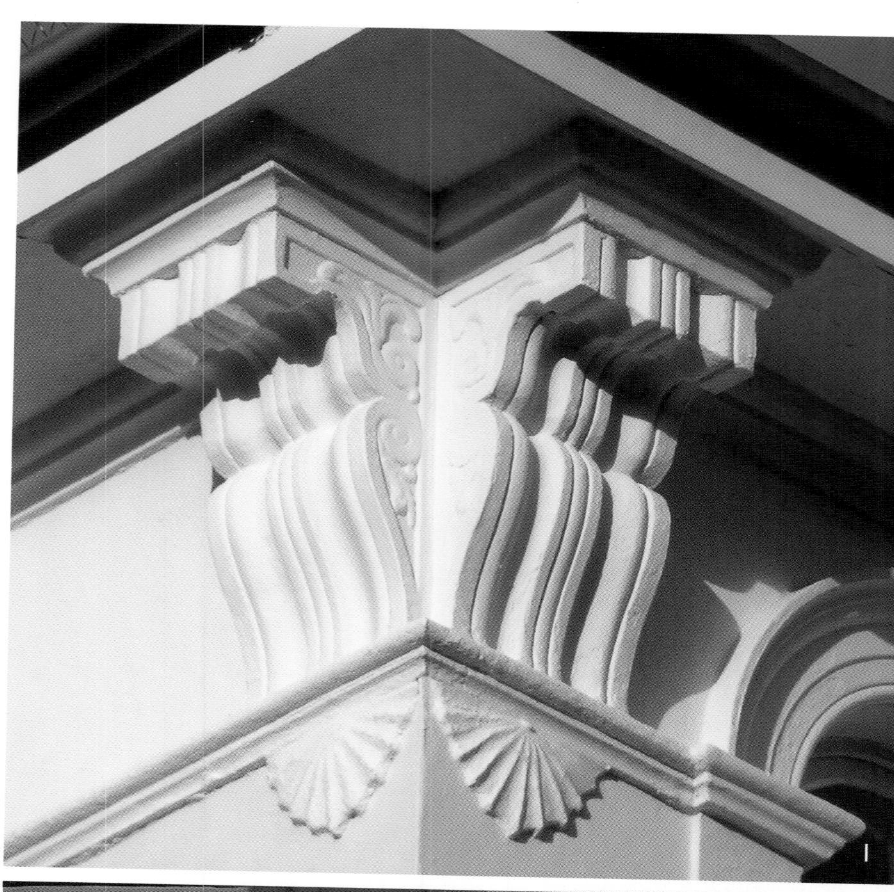
1

2

4

3

The majestic sweep of Eastbourne's select Victorian hotels in Grand Parade is terminated by the Queens Hotel designed by Henry Currey and opened in 1880. This was set forward of the main building line as a visual barrier to divide the high class hotels from the boarding houses to the east of the town. Eastbourne's stucco façades were carried out by specialist plasterers such as James Hookham who displayed an elegant miniature temple built of plasterwork at the Workmen's Exhibition in Devonshire Park in 1885. This was a popular feature in the Park until fairly recently when it was taken down as it was thought to be structurally unsafe.

this page: Trinity Trees

facing page:
1, 2 & 3. Trinity Trees
4. College Road

1

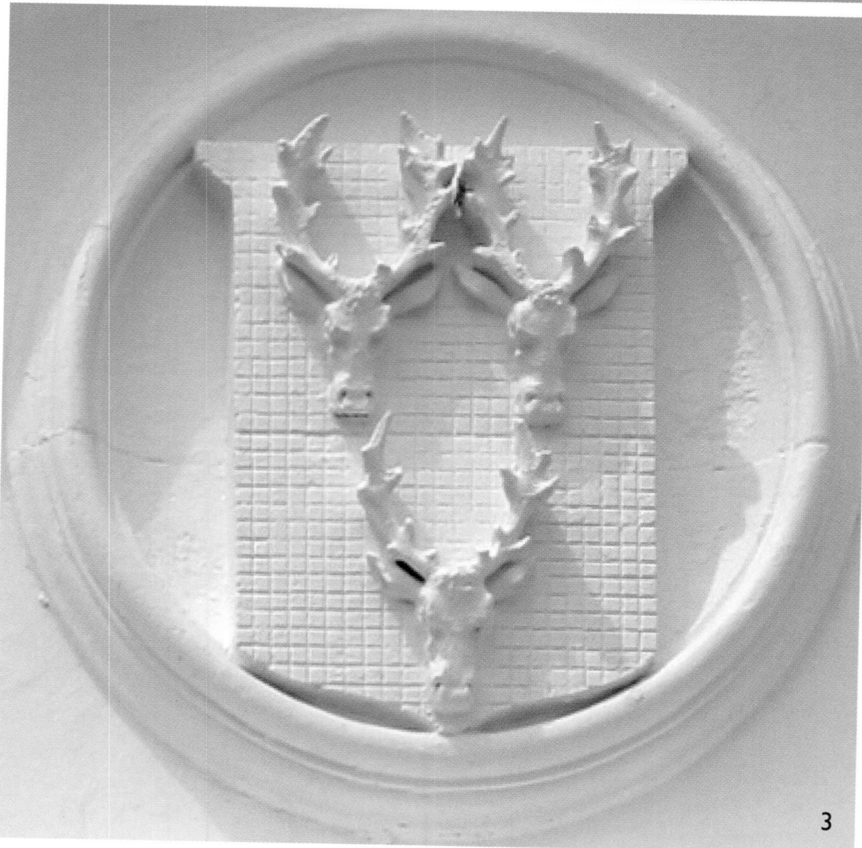
3

2

On the Lansdowne Hotel gatepost is a shield commemorating the Eastbourne Hunt which used to congregate on the Western Lawns before sallying forth on to the Downs. Other symbolic motifs found in the town are the head of Neptune with seaweed hair on the Chatsworth Hotel, the anchor of the former Anchor Hotel (now the Eastbourne Riviera Hotel) and the three stags of the Cavendish coat of arms on the Burlington Hotel.

this page: Chatsworth Hotel, Grand Parade

facing page:
1. Eastbourne Riviera Hotel, Marine Parade
2. Lansdowne Hotel, King Edward's Parade
3. Burlington Hotel, Grand Parade

pages 92/93: "Pargetting", South Street

09

decorative ironwork

Decorative cast ironwork was a notable feature of Victorian seaside resorts and we were fortunate enough to have our own ironmaster Ebenezer Morris, who had set up his foundry in Lewes in 1823 and opened a shop in Eastbourne in 1853 when the Earl of Burlington set about the development of the town. Later on his son joined him so that much of his ironwork displays the raised lettering E. Morris and Son. Cast ironwork is very much a product of the Victorian industrial age and perhaps represents an age of optimism, confidence and industry. The Victorians in particular had an extravagant passion for decoration and cast ironwork, which is made from molten iron poured into a mould, enabled an infinite variety of decoration to be formed.

Sussex was already known for producing cast iron from the Tudor period which gave rise to the great growth of the Wealden iron industry where, near Mayfield:

"Master Huggett and his man John,
They did cast the first can-non."

The casting process was complex and dependent on delicately carved patterns of fine grain timber which were used to form shapes in a sand mould from which the final castings were produced.

09

1

2

Some of the earliest railings in cast iron to be found in the town are at Cornfield Terrace, built in 1851, where simple spear-headed balusters were used. These were secured into the top horizontal rail with molten lead and set at their bases into a rendered plinth. Later on as the town developed, there were two main types of baluster employed in front of the terraces and villas. The first has a distinctive fleur-de-lys shaped head which was used throughout Eastbourne during the 1860s and 1870s. The Eastbourne fleur-de-lys railing head is a good example of the refined attention to detail adopted by the Victorians. The artistry of its designer has been fulfilled by the pattern-maker and reproduced with stunning effect in the heat of Ebenezer's Lewes foundry. Nothing can be added or taken away to improve this splendid object. The second type was the club head found in Cavendish Place and elsewhere. More elaborate railings were also used bearing the Morris name such as those in the fine bow-fronted terrace in Hartington Place. Ornamental railings were used not only to provide physical protection to the basements below street level but also to create a wonderful repetitive decorative element set against the stucco buildings behind.

1

2

As well as railings, a number of different cast iron elements were employed such as the Adam style cast iron panels in Cavendish Place which were based on Regency patterns including anthemion and rolling wave decoration. Cast iron was also used in providing ornamental enclosures for window planters on window sills, for supporting ironwork to canopies and in seats, gates and roof embellishments as we shall see later in this book. Sadly, much of Eastbourne's architectural cast iron decoration was removed during the Second World War as part of the war effort and has not yet been replaced.

page 95: South Street

this page:
1. Cavendish Place
2. Hartington Place

facing page:
1. Cornfield Terrace
2. Ceylon Place

page 98:
1. Cavendish Place
2. Howard Square
3. Hyde Gardens

page 99:
1. Grand Hotel, King Edward's Parade
2. Marine Parade
3. South Cliff
4 & 5. Eastbourne Heritage Centre, Carlisle Road

1

2

3

1
2
3
4
5

Decorative cast ironwork was used to provide structural columns and brackets in a number of public buildings including the Railway Station of 1886 for the London, Brighton and South Coast Railway. The Winter Garden at Eastbourne was designed by the Duke's architect Henry Currey, and is a building of cast iron columns combined with light tubular wrought iron roof ribs. As we will see in future chapters, the roofscape was ornamented by delicate cast ironwork. Eastbourne Pier is a handsome structure of cast iron screw piles supporting a wrought iron frame stretching out 1,000ft into the sea. It was designed by Eugenius Birch, the most prolific of pier builders and much of the casting was carried out by John Every of Lewes who was apprenticed in the Morris Foundry. The first pile of Eastbourne Pier was screwed into the hard clay of the seabed in 1866 but it was more than six years before the Pier was completed although it was opened on 13th June 1870. Originally built as a promenade pier and landing stage for paddle-steamers, it supported eight small kiosks and a bandstand at the head. The geometric cast iron seat back doubled up as a safety railing and the top rail also served a double use as a gas pipe for lighting on the original structure. The cast iron drinking fountain in Sea Houses Square with its entwined dolphin lantern support (shown on page 103) was donated in September 1865 by Mrs Elizabeth Curling who lived in Kent Lodge, Seaside Road (now Trinity Trees). It was originally located in the middle of the road in Seaside near the Leaf Hall, later moved to the junction of Langney Road, and finally to its present location when it was restored in 2000.

this page: Cast iron seat back, Eastbourne Pier

facing page:
1 & 2. Eastbourne Railway Station

page 102:
1. Shelter near Redoubt Fortress, Royal Parade
2. Seafront railing, Grand Parade
3. Cast iron seat back, Eastbourne Pier

page 103:
1. Drinking fountain, Sea Houses Square, Marine Parade
2 & 3. Shelter near Redoubt Fortress, Royal Parade

1
3
2

SAID
OF
THIS WATER
SHALL THIRST AGAIN
1
2
3

1

3

2

n the 20th century the modern movement in architecture dictated that "form must follow function" and the influential Viennese architect, Adolf Loos, announced that "ornament is a crime". This view still holds sway and it is extremely difficult for architects now to add decoration to their buildings without derision and criticism. There are signs however that this is gradually changing and architects have recently relished in fine detailing, if not decoration. The use of ornament is now not considered a criminal offence, but certainly cannot be employed without good reason. Panelled doors and decorative stair balustrades are no longer covered over with flush hardboard. It is now very acceptable and desirable to "put back the style", and to reinstate the lost ornamentation on Eastbourne's noble terraces and princely mansions. When this is done it is important that the correct details and materials are used following careful research.

this page:
1. Willingdon Road shopping parade
2. Eastbourne Bandstand, Grand Parade
3. Selwyn Road

facing page:
1. Eastbourne Town Hall, Grove Road
2. Eastbourne Heritage Centre, Carlisle Road
3. Baptist Chapel, Grove Road

10

terracotta

The sight of deep red terracotta on a fine sunny day can even supplant the delight given by Eastbourne's Victorian red brickwork. Terracotta is the Italian for "baked earth" and is closely related to brick, but using finer clay. It was popular briefly in the 16th century but came back into its own in the second half of the 19th century when it was widely used in many important buildings such as the Albert Hall. In domestic vernacular architecture, the material was often used to date a building or to display the initials of the owner such as the fine examples in Staveley Road (shown above) and Old Wish Road.

It is for good reason that the superb terracotta Catherine wheel window at the west end of All Souls Church has been reserved for our front cover. This church was consecrated in 1882, paid for by Lady Victoria Wellesley (great-niece of the Duke of Wellington), and deliberately located in the east side of the town to serve the working community. It is a striking example by architects Parr and Strong of the Victorian passion for reviving previous architectural styles – in this case, Lombardo-Byzantine architecture.

10

page 107: Staveley Road

facing page: Eastbourne College, Old Wish Road

this page:
1. Blackwater Road
2. Silverdale Road

1
2
3
4

A variation of terracotta is called faience (from the Italian town of Faenza). This is a glazed form of terracotta which was developed towards the end of the Victorian period. A good example is the green capital top on the corner of Furness Road and South Street which once housed Elliott's grocery store, the Fortnum's of Eastbourne. Faience was also widely used in the 1920s and 30s and we have splendid examples in the Grand Parade shelters, the former Luxor/ABC Cinema in Pevensey Road and in the neo-Grecian style Eastbourne Bandstand which was built in 1935 by Leslie Roseveare, Borough Architect, based on concept designs by John D. Clarke and A. F. Worsfield. The structure was originally faced in "ceramic marble" produced by Carter Stabler & Adams of Poole, Dorset.

this page: Corner of South Street and Furness Road

facing page:
1. Stafford House, Southfields Road
2. Eastbourne College, Old Wish Road
3. Grassington Road
4. Blackwater Road

1
2
4
3

this page: Highly glazed faience, Eastbourne Bandstand, Grand Parade

facing page:
1. Promenade shelter, Grand Parade
2. Former Luxor/ABC Cinema, Pevensey Road
3. Eastbourne Bandstand, Grand Parade
4. Premier Inn, Terminus Road

stained glass

Eastbourne displays a great variety of stained glass in its buildings, not only in churches but also in its civic buildings and domestic architecture. The medieval art of creating this glass was revived by the Victorians as the art had been largely lost so that in the 16th century, painted glass had become the norm such as carried out by German and Flemish glaziers at King's College Chapel, Cambridge. The Victorian revival of the medieval stained glass tradition meant that the early techniques had to be rediscovered. The stained glass is produced by various techniques. Yellow is the only glass obtained by true "staining" where silver nitrate is painted onto the surface and then absorbed into the glass by firing in a kiln. Pot metal glass has the colour right through the glass but where lighter colours are required such as in reds, thin glass is bonded to the clear glass to prevent the colour being too deep. Much of the glass is painted on the inside with special brushes for shading known as "matting" and a thinner brush for the line work. The fine detail of the hair shown in the illustration above of St Peter at St Saviour's Church is formed in this way. In order to fuse the paint or stain to the glass, it is fired in a kiln with a temperature raised to between 1100 and 1300 degrees Fahrenheit at which point the glass becomes soft thus allowing the paint to sink into its surface.

11

1

2

Glass is usually held in lead cames which are lead strips manufactured to form an H section surrounded by a putty known as glazing cement. The illustrations from St Saviour's Church are of glazing by Clayton and Bell. The partnership of A. R. Clayton and Alfred Bell was formed whilst they were working at the large architectural practice of Gilbert Scott where they formed friendships with George Edmund Street, the architect for St Saviour's. Eventually the firm employed about 300 people in their studio producing work all over England from the great cathedrals to parish churches. In All Saints Convalescent Hospital Chapel of 1874, the stunning glass by John Hardman Powell creates a work of shimmering harmony where seraphim with their red hot chilli-pepper wings soar heavenwards (shown on facing page).

this page: St Saviour's Church, South Street:
1. St George (c275-303), patron Saint of England, slaying the dragon
2. St Edwin (c586-633), King of Northumbria, holding a model of York Minster which he founded

facing page: Seraphim by John Hardman Powell, East window, All Saints Convalescent Hospital Chapel, Darley Road

I

ɛx Downs College, three fine windows removed
former art college building in St Anne's Road,
ne Town Hall various coats of arms are depicted
riumphal staircase. The arms of the Cavendish
d a serpent, the Eastbourne coat of arms include
seahorse crest (shown on page 121). The Town
86 at a cost of £40,000 and the stained glass
etailed finish with marble columns and pilasters,
narble mosaic floors with Minton tiles and the
oinery.

this page: Eversley House, Sussex Downs College, Cross Levels Way, detail

facing page:
1. University of Brighton, Trevin Towers, Gaudick Road
2. Eversley House, Sussex Downs College, Cross Levels Way

1

2

is page: Eastbourne Town Hall, Grove Road:
Cavendish coat of arms
Eastbourne coat of arms

cing page: The Death, Assumption and Coronation of the Blessed Virgin Mary,
ibune Gallery, All Saints Convalescent Hospital Chapel, Darley Road

The examples on these pages are from buildings where besides the stained glass panels are geometric shapes formed by the lead cames. The round bull's eye" panes are formed from the centre of a piece of semi-molten glass which is spun by hand during crown glass manufacture.

this page:
1. Former Tally Ho public house, Church Street
2. St John's Road

facing page: Granville Road

12

balconies, canopies and shelters

Many of Eastbourne's Victorian buildings are enhanced by canopies and balconies which take a variety of forms as shown in the photographs. Most employ the skill of the master carpenter with turned balusters, carved pilasters and repetitive ornamental fretwork which catches the light and throws shadows against the buildings and windows against which they are placed. Victorian ladies loved to be outside but not in direct sunlight and old photographs show that the parasol was as common a feature in the sunshine as the umbrella was in wet weather. At the Kings Arms in Seaside the first floor canopy is more like a gallery and is supported on stone brackets and covered with a curved zinc roof with a fretted timber fascia reminiscent of a railway station. Within the room behind this, Don Cockel the British Commonwealth boxing champion, used to practice in the late 1950s. At All Saints Convalescent Hospital (1867-69) we find Henry Woodyer creating a rich Bath stone balcony that could come straight out of a Venetian Palazzo fronting the Grand Canal with a quatrefoil balustrade fitted in with an Early English arcade of round columns with bell capitals and chamfered Gothic arches (shown on page 127).

12

Sadly, many Victorian buildings have lost their original canopies (e.g. Hyde Gardens, the Queens Hotel and the bow-fronted terrace at the sea end of Terminus Road which was originally called Victoria Place) but at the late Victorian East Beach Hotel there is an outstanding example of a canopy with delicate pierced ornamental fretwork which adds so much to the fun and jollity of the seaside. An original surviving zinc canopy adorns the bay window of Bolton House with its lacy pierced frieze throwing delightful shadows against the glass (shown on page 129).

page 125: Kings Arms, Seaside

this page: Bishop Carey, Gaudick Road

facing page: All Saints Convalescent Hospital, Darley Road

The Regency style terraces in Cavendish Place (which is in fact from the 1850s) also displays circular zinc canopies and there is one example where the fascia is cut and pierced. Originally all the canopies were painted with different coloured stripes. Some canopies were supported on cast iron decorative panel columns with Regency style brackets with anthemions and scrollwork such as those found in Upperton Gardens and Pevensey Road. Glazed canopies were also used to decorative effect in providing a weatherproof cover over grand entrance steps to Victorian buildings (shown on page 132).

page 128: East Beach Hotel, Royal Parade

page 129:
1 & 2. Cavendish Place
3. Bolton House, Bolton Road
4. Pevensey Road
5. East Beach Hotel, Royal Parade

A number of seafront thatched shelters with their herringbone brickwork and decorative timber colonnades adorn the Western Parades from the Wish Tower to Holywell (shown on page 133). The three levels of walks and terraces were paid for by the 7th Duke, designed by Henry Currey and opened by the Prince of Wales on his royal visit on 30th June 1883. A much earlier 18th century shelter known as the Paradise Belvedere can be found tucked away below Paradise Drive at the western edge of the golf links. This finely knapped and squared flint folly has a central pedimented arch and two wings with arched niches and is thought to have been erected by Lord Wilmington in 1739.

In the Manor Gardens stands a late 18th century Gothick gazebo known as "The Hermitage", polygonal in shape with a thatched roof and "very pretty in its wedding-cake colouring" (Antram & Pevsner). This gazebo along with Gildredge Manor House, is thought to have been designed by Sir Robert Taylor.

this page: The Hermitage, Manor Gardens
facing page: Paradise Belvedere, Paradise Drive

1

3

2

this page: Western Parade shelter

facing page:
1. Spencer Road
2. Pevensey Road
3. Bolsover Road

13

decorative tiling and mosaics

Once again, the Victorians looked to the medieval past for their inspiration in the revival of ornamental tiling and mosaics. In medieval times, encaustic tiles were made for cathedrals and churches. These are ceramic tiles in which the pattern on the surface is not a product of the glaze but of different colours of clay inlaid into the body of the tile so that the design remains the same as the tile is worn down. These tiles may be glazed or unglazed and the inlay may be as shallow as an 1/8 or as deep as 1/4 of an inch. Augustus Welby Pugin the great Victorian medieval revivalist who carried out the decoration and ornamentation of the Palace of Westminster in the 1850s used encaustic tiles produced by Mintons China Works of Stoke-on-Trent. Herbert Minton was one of the outstanding entrepreneurs of the 19th century and he worked closely with Pugin, Sir Henry Cole and even Prince Albert to produce their flamboyant designs. In Eastbourne, we have examples of Minton tiles in domestic buildings, in the Victorian churches such as St Saviour's, Christ Church and All Souls, in the All Saints Convalescent Hospital and in the Town Hall. St Saviour's Church was dedicated in 1867 and designed by George Edmund Street. He worked closely with the famous firm of Clayton and Bell on a set of fine mosaics.

13

page 135: South Street

this page: "The Sower and his Seed", St Saviour's Church, South Street

facing page: "Blessed Virgin Mary", St Saviour's Church, South Street

The glass mosaic in the Chancel depicting the "Blessed Virgin Mary" was made and assembled by the famous firm of Salviati in Murano, Venice. This partnership had already carried out the extensive mosaic work to the towering Albert Memorial in Hyde Park. This is one of a set of shimmering mosaics in an embellished arcade supported on highly polished Sussex and other marble shafts. Another fine mosaic by Clayton and Bell is found at the west end of the church depicting the parable "The Sower and his Seed". This time, the mosaic is made up of ceramic pieces.

this page:
1. Encaustic tiles, St Saviour's Church, South Street
2. Fusciardi's Ice Cream Parlour, Marine Parade

facing page:
1. Mosaic, Eastbourne Town Hall, Grove Road
2. Carlisle Road
3. Encaustic tiles, Lady Chapel, St Saviour's Church, South Street

SALVE
1
3
2

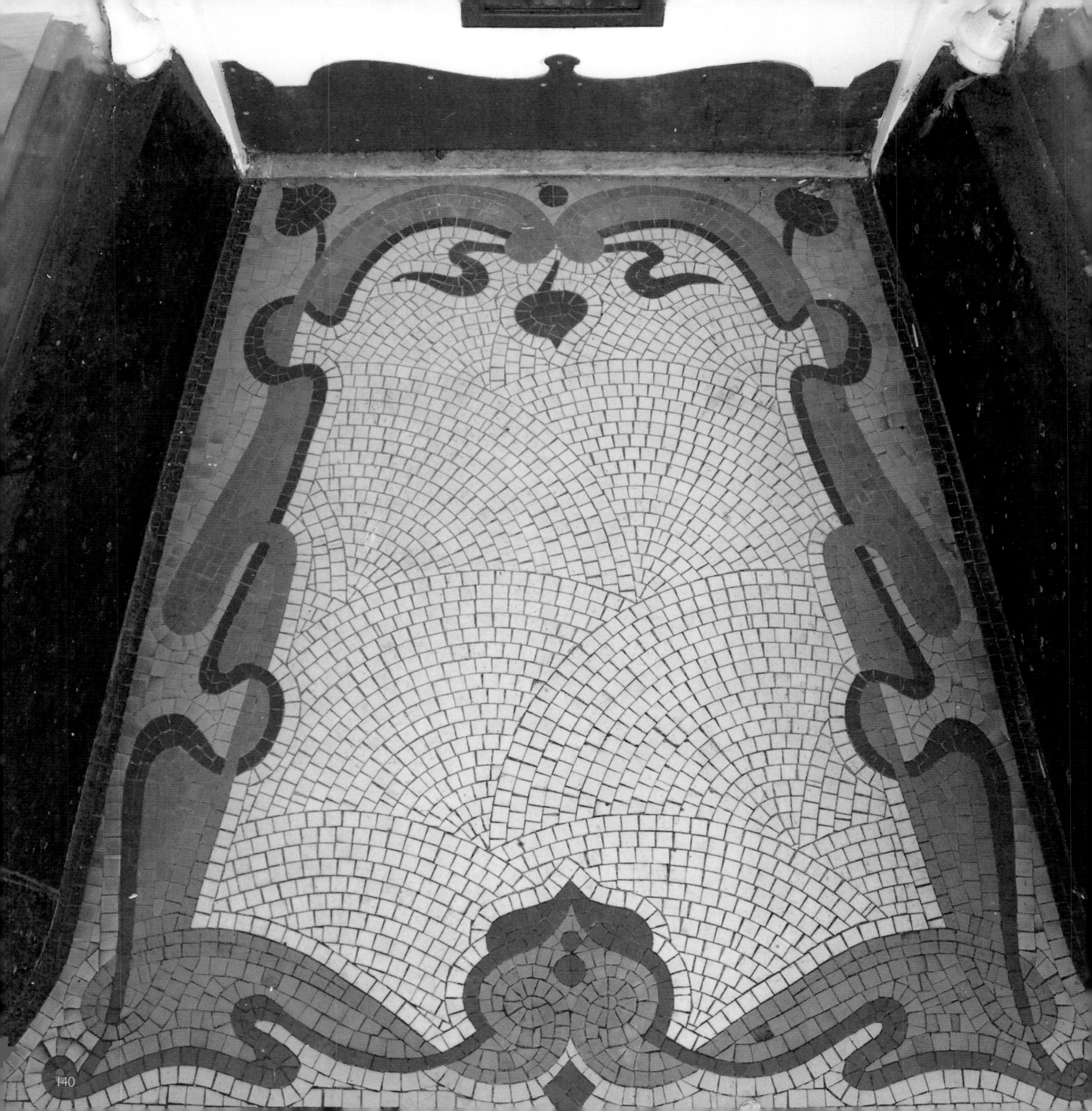

this page: Former Hudson's shop, Compton Street, where several shops in Grand Hotel Buildings contain the names of their former business owners

facing page: Cornfield Road

14

plaques

A number of interesting plaques are displayed around the town. These are attached to buildings and monuments with the intention of commemorating events or providing the viewer with information. On this page is the Eastbourne coat of arms carved in Portland stone at Eastbourne Town Hall which was opened by the Mayor on 20th October 1886 to the strains of the Hallelujah Chorus. The coat of arms contains two stags from the Cavendish arms and a rather stiff looking seahorse for its crest together with the inscription "Meliora Sequimur" meaning "let us follow better things".

14

EASTBOURNE WATER WORKS COMPANY
INCORPORATED BY ACTS OF PARLIAMENT 1858-1881,
CHAIRMAN
HIS GRACE THE DUKE OF DEVONSHIRE.
THIS STONE
WAS LAID BY MISS ADA CURREY.
ON THE 3RD DAY OF OCTOBER 1881.
HENRY CURREY. F. R. I. B. A. ARCHITECT.
GEORGE A. WALLIS M. I. C. E. ENGINEER.

This page shows the foundation stone of Bedfordwell Pumping Station for the Eastbourne Waterworks Company. The building was designed by Henry Currey and the stone laid by his daughter Ada. The engineer was George Ambrose Wallis. This building was later opened by the Prince and Princess of Wales during their royal visit in June 1883. Another plaque featuring Mr Wallis is found set in the promenade edging near the Redoubt to commemorate the eastward extension of the Royal Parade sea wall in 1884. The infant school plaque is from the school building in Meads Road paid for by the 2nd Earl of Burlington's wife Blanche, erected mainly for the benefit of children on the Compton Estate. The Countess died in 1840 aged 28.

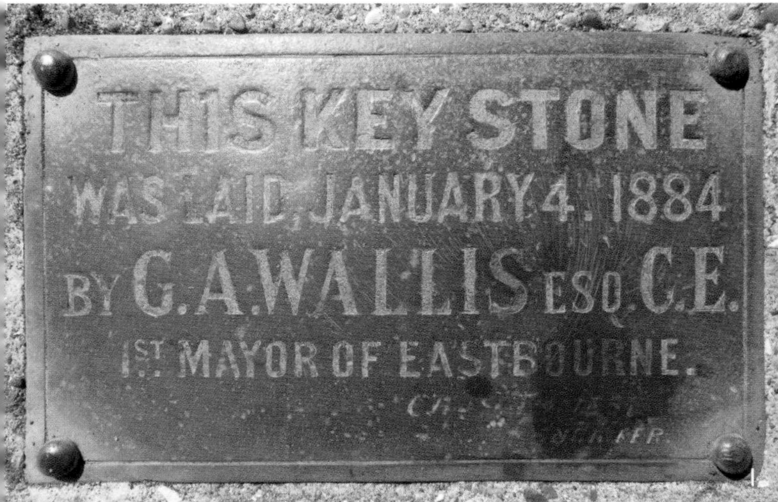
THIS KEY STONE
WAS LAID, JANUARY 4. 1884
BY G.A.WALLIS ESQ. C.E.
1ST MAYOR OF EASTBOURNE.

page 143: Eastbourne Town Hall, Grove Road

this page:
1. Royal Parade, promenade near the Redoubt Fortress
2. Plaque removed from the former Princess Alice Hospital, Carew Road
3. St Michael and All Angels Church, Willingdon Road

facing page: Bedfordwell Pumping Station, Bedfordwell Road

1

3

Sacred to the Memory of HENRY LUSHINGTON
Eldeſt Son of HENRY LUSHINGTON D.D. Vicar of this Pariſh And MARY his Wife
Whoſe singular Merits & as singular Sufferings Cannot fail of endearing Him to ye lateſt Poſterity.
At ye Age of Sixteen in ye Year 1754 He embarqued for BENGAL in ye Service of ye India Company, & by
attaining a perfect Knowledge of the Perſian Language made Himſelf eſſentially uſeful. It is difficult to
determine whether He excelled more in a civil or a military Capacity. His Activity in Both
recommended Him to the Notice & Eſteem of Lord CLIVE: Whom with equal Credit to
Himſelf & Satisfaction to his Patron He served in the different Characters of
Secretary, Interpreter & Commiſſary. In ye Year 1756, by a melancholy Revolution, He was with Others
to ye Amount of 146 forced into a Dungeon at CALCUTTA so small that 23 only eſcaped Suffocation. He was
One of ye Survivors, but reserved for greater Miſery, for by a Subſequent Revolution in the Year 1763 He was
with 200 more taken Prisoners at PATNA, & after a tedious Confinement being singled out with JOHN ELLIS &
WILLIAM HAY Esqrs was by the Order of the Nabob COSSIM ALLY KAWN & under ye Direction of One SOMEROO
an Apoſtate European, deliberately & inhumanly murdered: But while ye Seapoys were performing their savage Office
on ye firſt mentiond Gentleman, fired with a generous Indignation at the Diſtreſs of his Friend,
He ruſhed upon his Aſſaſſins unarmed, & seizing One of their Scimitars killed Three of Them & wounded Two Others,
till at length oppreſſed with Numbers He greatly fell.
His private Character was perfectly consistent with his publick One. The amiable
Sweetneſs of his Diſpoſition attached Men of ye worthiest Note to Him. the Integrity of his Heart fixed them ever
firm to his Interests. As a Son, He was One of the most kind & dutiful, as a Brother ye most affectionate.
His Generosity towards his Family was such as hardly to be Equal'd, his Circumstances and his Age
conſider'd, scarce to be exceeded. In short He lived & died an Honour to his Name, his Friends & his Country.
His Race was short (being only 26 Years of Age when He died) but truly glorious.
The rising Generation must admire, May They imitate, so Bright an Example!
His Parents have erected this Monument, as a lasting Testimony
of their Affliction & of his Virtues.

2

1

2

3

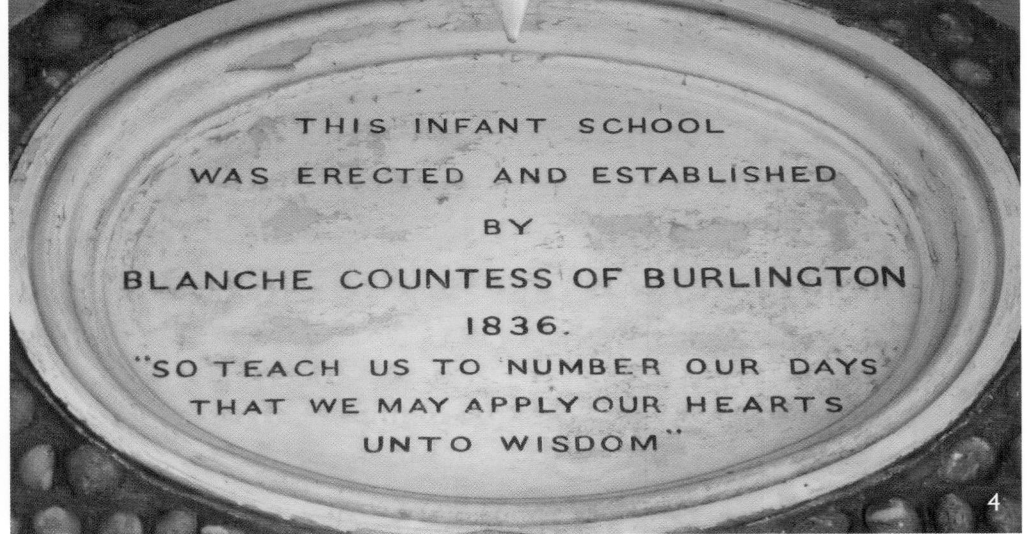

THIS INFANT SCHOOL
WAS ERECTED AND ESTABLISHED
BY
BLANCHE COUNTESS OF BURLINGTON
1836.
"SO TEACH US TO NUMBER OUR DAYS
THAT WE MAY APPLY OUR HEARTS
UNTO WISDOM"

4

Other plaques displayed on these pages include the coat of arms of the London, Brighton and South Coast Railway on Eastbourne Station of 1886, the arms of Eastbourne College and various tablets from the biblical inscription on the drinking fountain in Sea Houses Square to a plate on the side of the statue to the Royal Sussex Regiment. An interesting detail is the lead rainwater head at Gildredge Manor which has the raised initials HLM for the Reverend Dr Henry and Mary Lushington for whom the house was built in around 1776 as previously mentioned. Lushington's son, also Henry, was one of the survivors of the Black Hole of Calcutta, and his memorial by Sir Robert Taylor (who may have designed Gildredge Manor) can be found in the south aisle of St Mary's Church.

this page:
1. Gildredge Manor, Borough Lane
2. Blue plaque, Milnthorpe Road
3. Blue plaque, Lushington Road
4. Former school, Meads Road

facing page:
1. Eastbourne Railway Station
2. Henry Lushington Memorial, St Mary's Church, Church Street
3. Eastbourne College

page 148:
1. Former Ship Inn, Meads Street
2. War Memorial, Town Centre
3. Hodeslea, Buxton Road

page: 149
1. Western Parades chalets
2. Baptist Chapel, Grove Road
3. Hurst Arms, Willingdon Road
4. Leaf Hall, Seaside

18
G. LUCK.
97
THESE PREMISES ARE
ERECTED ON THE SITE
OF THE OLD SHIP INN
WHICH WAS BUILT ABOUT
A. D. 1600.

1

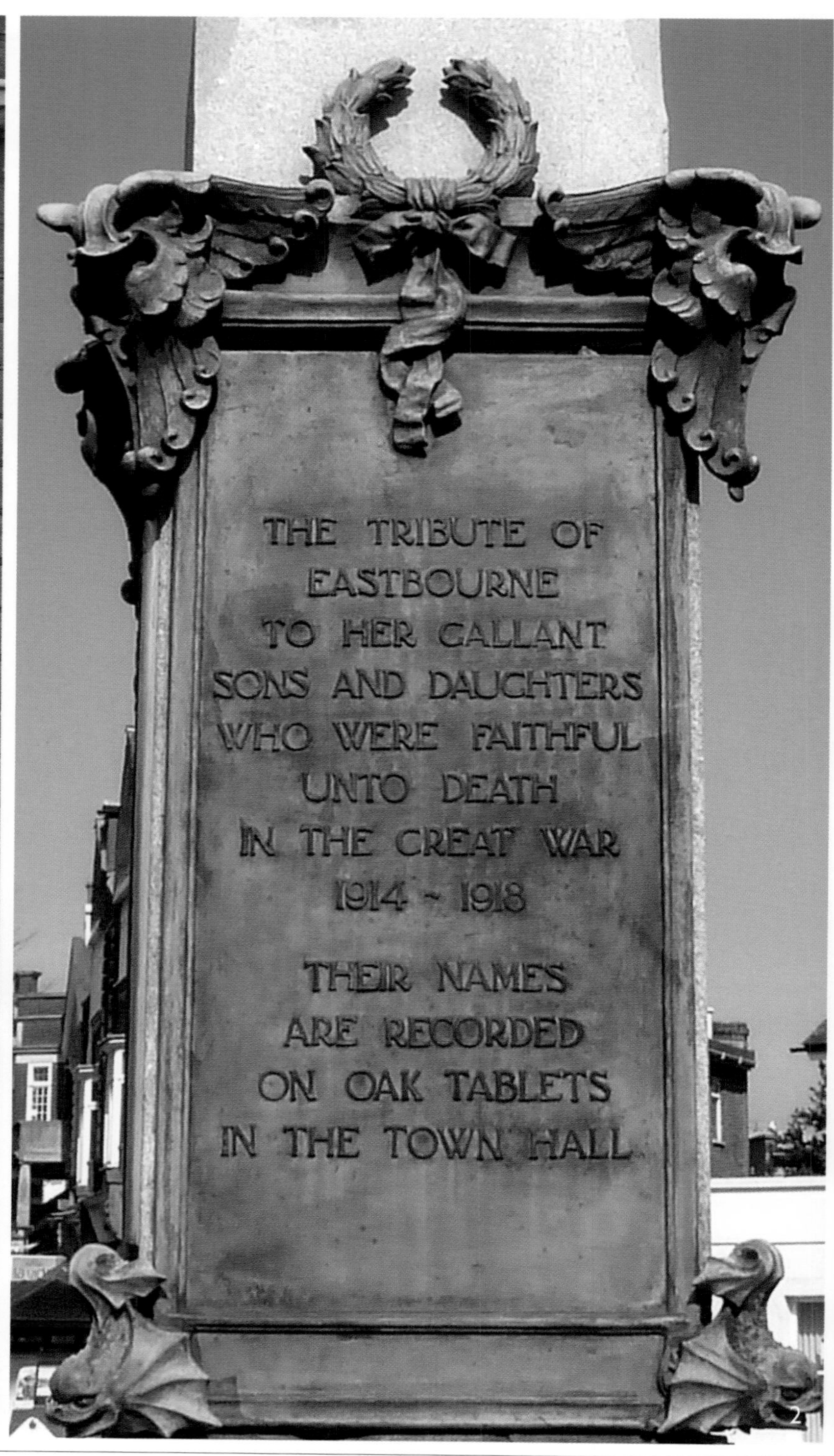
THE TRIBUTE OF
EASTBOURNE
TO HER GALLANT
SONS AND DAUGHTERS
WHO WERE FAITHFUL
UNTO DEATH
IN THE GREAT WAR
1914 ~ 1918
THEIR NAMES
ARE RECORDED
ON OAK TABLETS
IN THE TOWN HALL

2

THIS HOUSE "HODESLEA" WAS BUILT BY
THOMAS HENRY HUXLEY. F.R.S.
1890 HE LIVED HERE 1895

3

THIS CHALET WAS USED BY THEIR MAJESTIES KING GEORGE V
AND QUEEN MARY IN THE MONTH OF MARCH. 1935

1

4

GROVE ROAD
BAPTISTS CHAPEL
ERECTED A D 1881
J. A. SKINNER. BUILDER.

2

3

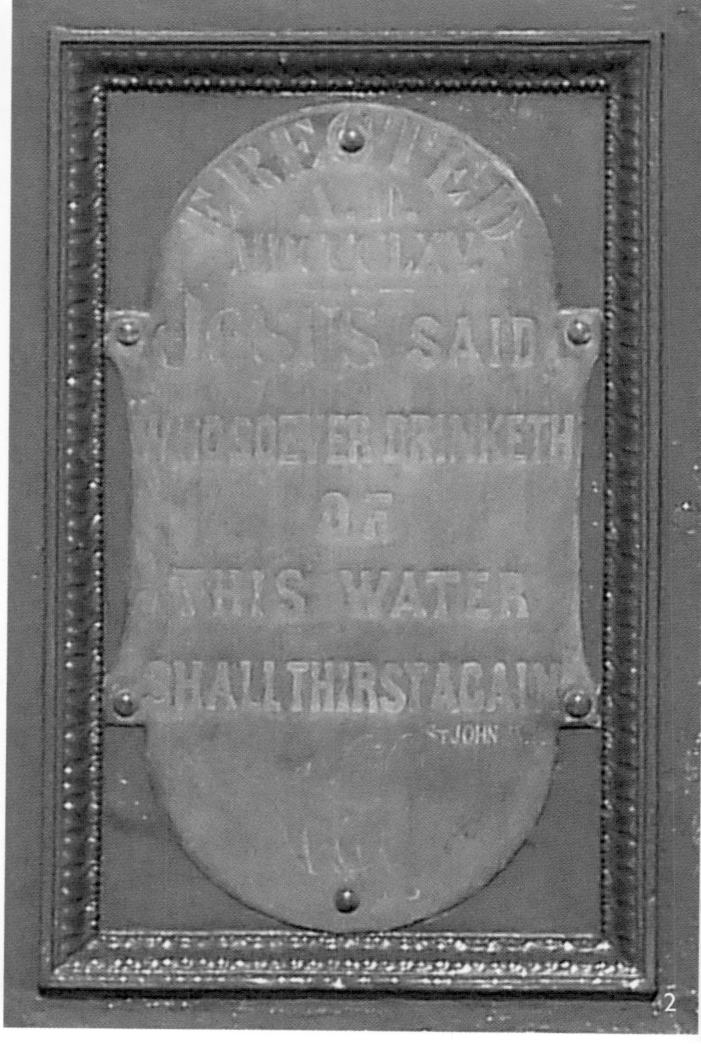

At the Eastbourne Bandstand a memorial commemorates John Wesley Woodward, one of the eight hero musicians who went down playing as the *Titanic* sank on 15th April 1912. Woodward was a 'cello player who had lived in Eastbourne for many years, playing in the Eastbourne Municipal Orchestra, the Duke of Devonshire's Orchestra and the Grand Hotel Orchestra. The memorial was designed by Eastbourne's prominent sculptor Charles Godfrey Garrard.

this page:
1. John Wesley Woodward Memorial, Eastbourne Bandstand
2. Drinking Fountain, Sea Houses Square

facing page:
1. Eastbourne Bandstand, Grand Parade
2. 2nd Battalion of the Royal Sussex Regiment Memorial, Grand Parade
3. Wish Tower (Martello Tower number 73), King Edward's Parade

MELIORA SEQUIMUR
OPENED BY
THE RT. HON
LORD LECONFIELD
G.C.V.O.
LORD LIEUTENANT
OF SUSSEX
ON THE
5TH AUGUST 1935
COUNCILLOR
MISS THORNTON J.P.
MAYOR
ALDERMAN
Lt.Col. ROLAND GWYNNE
D.S.O., D.L., J.P.
CHAIRMAN OF THE COMMITTEE
LESLIE ROSEVEARE
M. INST. C.E.
BOROUGH ENGINEER
LXXII

15

clocks and sundials

In Eastbourne, there are many clock faces that adorn churches, schools and public buildings. Of particular note is the clock face of All Souls Church which, as we have already seen, was designed by Parr and Strong and paid for by Lady Victoria Wellesley, great-niece of the Duke of Wellington. The 83ft Lombardo-Byzantine free-standing campanile is constructed of yellow stock brickwork laid in Flemish Bond and embellished with red brick and terracotta dressings. The four clock faces of the 130ft high Town Hall tower of 1886 were not put in until 1892 when the cheerful Westminster chimes were installed by the famous clockmakers Gillett & Johnston of Croydon. This firm also installed the clocks at the Leaf Hall and at Eastbourne Railway Station. There are very few sundials in Eastbourne. The earliest sundial illustrated is on the 18th century farmhouse of Meads Place on the corner of Meads Road and Gaudick Road (shown on page 155). This building now surrounded by Victorian Meads is constructed of red bricks with glazed grey headers and the surround of the sundial is in a Baroque style with a broken pediment. The other example of a sundial is in flamboyant Jacobean style in Fairfield Road at Parkholme, dated 1897. Peeping out from under the semi-circular pediment is a carved blazing sun which appears to have cast a good strong shadow indicating that our photographer was up at 8.30 in the morning!

15

age 153: St Mary's Church, Church Street

acing page: All Souls Church, Susans Road

his page:
. Meads Place, Gaudick Road
. Parkholme, Fairfield Road

page 156:
1. Eastbourne College, Blackwater Road
2. Leaf Hall, Seaside
3. Eastbourne Railway Station
4. Eastbourne Pier entrance

page 157: EastbourneTown Hall, Grove Road

1

2

4

3

16

signage

Beautifully executed lettering, as well as being informative, can embellish a building and stylistically give a good clue to the date of the property. The photographs on these pages display an imaginative use of lettering from the free Art Nouveau style at the Saffrons and the former Tally Ho public house through to the 1960s barbers sign at the Railway Station. Some signs have been overtaken by time and events so that, for example, the 1911 shield carrying the letters BC on Debenhams Store give a clue to the fact that in the Edwardian period, this was Eastbourne's finest department store – Bobby and Company – the Harrods of Eastbourne. Here at various times of the day, one could listen to a trio of fine musicians playing in the restaurant under a glazed domed ceiling. At Calverley Road is one of only a handful of original enamelled blue and white street signs which once were commonplace throughout Victorian Eastbourne.

16

AYMOND.GRANGE
12
1

THE
WEDGE
2

SUMMERDOWN
C&B
IPWICK.ROAD.
5

GENTS
HAIRDRESSER
3

PLEASE ADJUST YOUR DRESS
4

1

2

3

page 159: The Saffrons, Meads Road

this page:
1. Burlington Hotel, Grand Parade
2. The Congress Theatre, Carlisle Road
3. Bibendum, Grange Road

facing page:
1. Aymond Grange, Dittons Road
2. The Wedge, Spencer Road
3. Gentlemen's hairdresser, Eastbourne Railway Station
4. Former gentlemen's urinal, Hampden Park
5. Corner of Church Street and Victoria Drive

page 162:
1. Former Bobby & Co (now Debenhams), Terminus Road
2. Haine and Son, South Street
3. Lamb Inn, Church Street,

page 163:
1. Former Tally Ho public house, Church Street
2. Central Buildings, Seaside Road
3. Post Office, Upperton Road
4. Enamel street sign, Calverley Road,
5. Enamel street sign, Granville Road
6. South Cliff Tower, Bolsover Road

B
1911.
C
HAINE & SON
The Lamb
A D 1180
HARVEYS OF LEWES
1
2
3

CALVERLEY ROAD

4

GRANVILLE ROAD

5

SOUTH
CLIFF
TOWER

6

17

chimneys

A relative of mine Mrs E. A. Jacob in 1933 once wrote a booklet entitled *Pots and Personalities* depicting the various types of chimney pot with some delightful illustrations. I think she had a point because chimneys and chimney pots have character and can add greatly to the "personality" of a building. For example, the photograph on this page shows a surprising marriage, where two separate buildings with a wall running between them are joined high above with an ornamental brick arch. A complete piece of whimsy but a delight to behold. The bricklayer and stonemason must have enjoyed greatly the challenge of constructing the chimneys shown on the following pages.

17

2

Chimneys and pots break the skyline and can provide expressive statements about each building. In Dittons Road we have pair of Tudor-style chimneys joined at the head. The stacks are turned at 45° and joined at the corner at the Prince Albert public house in High Street, Old Town. We have quite a serious composition at the Town Hall where each stack is seemingly supported on a colonnade of Ionic capitals topped with a heavy dentilled triangular pediment. Either side of the stack a pair of ornamental urns stand guard. The simple old stack on the Lamb Inn supports "father and son" pots.

page 165: Saffrons Road

this page:
1. Eastbourne Town Hall, Grove Road
2. Lamb Inn, High Street

facing page:
1. Dittons Road
2. Prince Albert public house, High Street

In Furness Road, the stack gradually diminishes with courses of brickwork laid at right angles to the slopes called "tumbling". The "granny" pots of the Devonshire Park Hotel each have a bonnet whereas the zany combined stacks of the Hydro Hotel are surmounted with a series of Bart Simpson look-alikes following their parents. The Art Nouveau Edwardian stacks in the Saffrons Road are cleverly detailed. The octagonal buff chimney pots in St John's Road are formed into shafts of medieval nail head decoration whereas the last two stacks shown at the Grand Hotel and on Eastbourne Railway Station beautifully complement the architecture of the building with their correctly proportioned Classical copings with "cyma recta" mouldings.

this page:
1. Furness Road
2. Devonshire Park Hotel, Carlisle Road

facing page: St John's Road

this page: Hydro Hotel, Mount Road

facing page:
1. Eastbourne Town Hall, Grove Road
2. Saffrons Road
3. Eastbourne Railway Station
4. Grand Hotel, King Edward's Parade

1

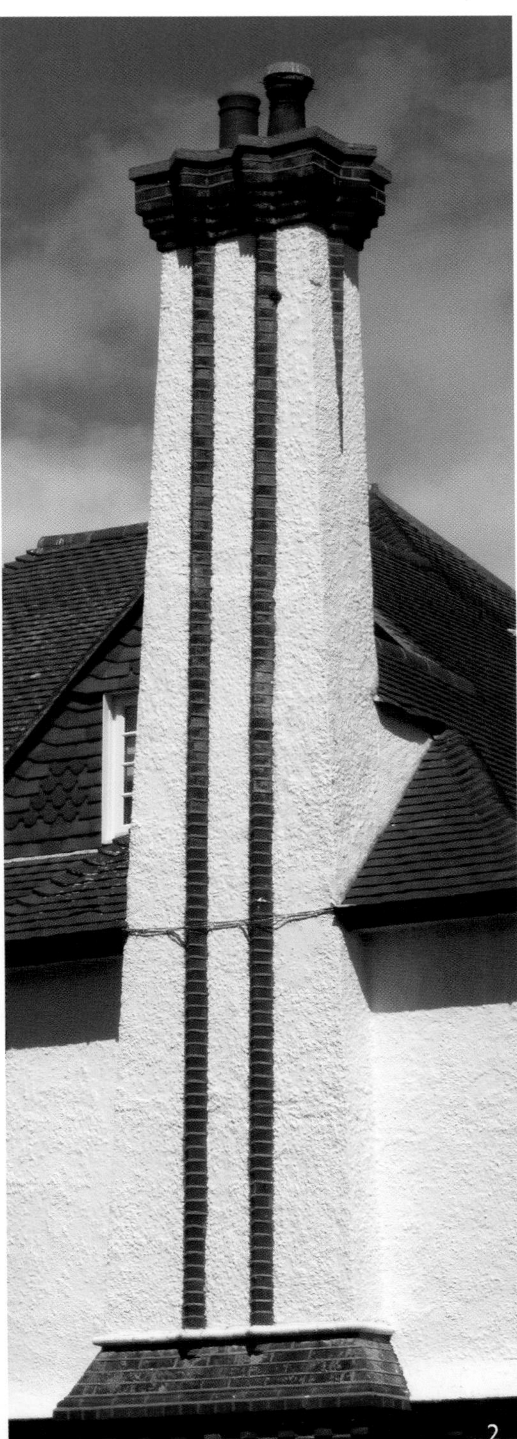
2

3

4

18

roof embellishments

In this, and the following sections of the book, we move from what might be described as careful detailing of functional parts of a building to decorative features that have been added to enrich the skyline and add decoration for its own sake to "top out" a building. Often they are pure flights of fancy but usually, when this decoration is removed, the buildings are much the poorer for it. For this reason the Eastbourne Society has been campaigning for many years to add back missing detail from roofs. One successful campaign was at Eastbourne Railway Station where the missing ironwork was reinstated recently. The present Station was completed in 1886 and is the fourth on the site. It was designed by F. D. Bannister, engineer to the London, Brighton and South Coast Railway. This delightful Victorian building, a mixture of medieval and classical styles, is dominated by its fine clock tower (shown on page 193). It also has a French pavilion style roof supporting a flagpole and a Brighton turret which is shown here. This at one time lit the original double-height booking office. The repetitive Jacobean style motifs of the cast iron roof frieze can be found on many railway stations around the country.

18

The photographs on these pages show the cast iron decoration which has also been reinstated recently on the roof of the Winter Garden. These follow carefully the original designs from Walter Macfarlane & Co.'s original Victorian pattern books. The turret balustrading surrounds a decorative flagpole support system all based on original photographs. The roof frieze has repetitive husk and scroll decoration and terminates in flamboyant vanes. Winter Gardens began to spring up in fashionable resorts in the country during the late Victorian period. These were light and airy structures full of exotic vegetation even in the middle of winter. The Winter Garden was designed by the 7th Duke's architect, Henry Currey, and appears to be influenced by the work of Decimus Burton (the Palm House at Kew) and Joseph Paxton (the Crystal Palace) with whom Currey had previously worked. The Floral Hall at the lower level was opened in August 1875 as a banqueting hall and roller skating rink opened by Mr J. C. Plimpton, the inventor of the roller skate. The upper level pavilion was opened later in July 1876 as a concert hall and ballroom. On 18th July 1881, electric light illuminated the Floral Hall for the first time. The local press described the view from the grounds: "the Floral Hall – like a miniature Crystal Palace – wore the appearance of one of the enchanted palaces to be read of in the Arabian Nights".

page 173: Eastbourne Railway Station
this and facing page: The Winter Garden, Compton Street

this page:
1. Former Hermitage's Piano Store, 116 Terminus Road
2. Granville Road

facing page: Eastbourne Town Hall, Grove Road

On these pages, we have other examples of iron decoration including the cresting on the barrel-vaulted roof of the former Hermitage's Piano Store where Claude Debussy is believed to have bought a piano during his stay at the Grand Hotel. Also shown is a turret ridge panel in Granville Road with its pair of crown finials reminiscent of a French chateau, and the flamboyant corona supporting the flagpole at Eastbourne Town Hall of 1886. Ironwork (cast and wrought) was not the only material employed to embellish a roof.

On this page, we have the carved stone finials of the Town Hall and the medieval style fleur-de-lys employed on the pinnacles flanking the main spire at St Saviour's Church in South Street. For about 150 years the surviving urns on the parapet of Sussex Gardens have terminated the vista at the end of Bolton Road. The urns which stand 4ft 6in high are beautifully detailed with fluted lids, vermiculated bands, garlands and satirical Bacchus heads which, like the ancient Green Man, are part human, part vegetation – in this case, vine leaves. One of these urns has been donated to the Eastbourne Heritage Centre and can be studied in detail there.

this page:
1. Urn from Sussex Gardens at Eastbourne Heritage Centre
2. Sussex Gardens urn above shops in Terminus Road

facing page:
1. Eastbourne Town Hall, Grove Road
2. St Saviour's Church, South Street

1

2

Tiled roofs were embellished with a great variety of decorative ridge tiles and terracotta finials which took the form of dragons, beasts, camels, eagles and ball and leaf finials. These were cast in fine clay and then carefully finished by hand and many were made by Gravett's terracotta works of Burgess Hill. The maker's name is inscribed above the maiden's head on a finial in Meads Road (shown on facing page).

this page:
1. Saffrons Road
2. The Goffs

facing page:
1. Meads Road
2 & 3. Victoria Drive

Not only the roof ridge was embellished but often the roof gables were enlivened by decorative bargeboards, carved and fretted in a variety of shapes and styles as shown in the photographs. Bargeboards were often terminated at the apex with turned finials and pendants. This type of decoration was loved by the romantic Victorians who delighted in the various shapes and the ever-changing interplay of light and shade they produced across the building.

this page:
1. Granville Road
2. Selwyn Road

facing page:
1. Meads Road
2. South Street

1
2

The roof embellishments on this page show a variety of finial terminations to turrets and cupolas which will be explored further in the next chapter. Shown is the gilded ball of the Town Hall clock tower, the oriental Indian-style dome and turret finials of The Buccaneer public house (built by the Devonshire Park and Baths Company in 1897 to the designs of Hastings architect, Henry Ward), and the striking Art Deco stainless steel finial which has recently been restored on the Eastbourne Bandstand of 1935. Within the constraints of the Modern Movement, the curvy roof of Grand Court of 1957 by Hugh Hubbard-Ford can also be said to have an embellished roof which echos the rolling sea that faces it.

this page:
1. Eastbourne Town Hall, Grove Road
2. Buccaneer public house, Compton Street

facing page:
1. Eastbourne Bandstand, Grand Parade
2. Grand Court, King Edward's Parade

1

2

19

turrets and towers

Eastbourne has a varied collection of turrets and towers and the earliest example shown here is the Wish Tower. This is tower number 73 of the 74 Martello Towers built along the coast between Aldeburgh in Suffolk and Seaford, (plaque "73" shown on page 151). They were constructed of local brickwork and built between 1805 and 1810 by the Royal Engineers to defend against the threat of French invasion. The name is derived from "Torre della Martella" in Corsica which had been spotted by the English during a campaign of 1794. Turrets were a good opportunity for architects and designers to add a flourish and focal point to their designs.

19

At the seaside, it was fairly easy to give free architectural expression and we have a variety of designs from the purely ornamental turrets on the 1901 games saloons on Eastbourne Pier with their fish-scale zinc shingles to the look-out turret of the Queens Hotel to the complicated Camera Obscura turret on Eastbourne Pier which topped the Pier Theatre built in 1899. The Camera Obscura made use of a large white painted dish housed in a darkened room under the dome upon which an image projected from a moving mirror in the revolving cupola (operated by a leather belt and cast iron windlass below) was enlarged via a large glass lens. At the time this was an almost unbelievable phenomenon to visitors who had never seen a moving colour image before (something which we now take totally for granted).

page 187: Wish Tower, King Edward's Parade

this page: Eastbourne Pier:
1. Gaming Saloon
2. Camera Obscura

facing page: Queens Hotel, Grand Parade

2

The octagonal Eastbourne Heritage Centre tower with its red Eastbourne tuck-pointed bricks set against carefully detailed Classical stucco work was built in 1880 as the manager's house and flag tower for the Devonshire Park and Baths Company. It is now home to the Eastbourne Society and is regularly open to the public, displaying the development of the town and its history. Public clocks provide a good excuse to construct towers such as those found at Eastbourne Town Hall (shown on page 157), Eastbourne Railway Station and the Leaf Hall.

this page:
1. Eastbourne Heritage Centre, Carlisle Road
2. Leaf Hall, Seaside

facing page: St Andrew's Prep, Meads Street

1

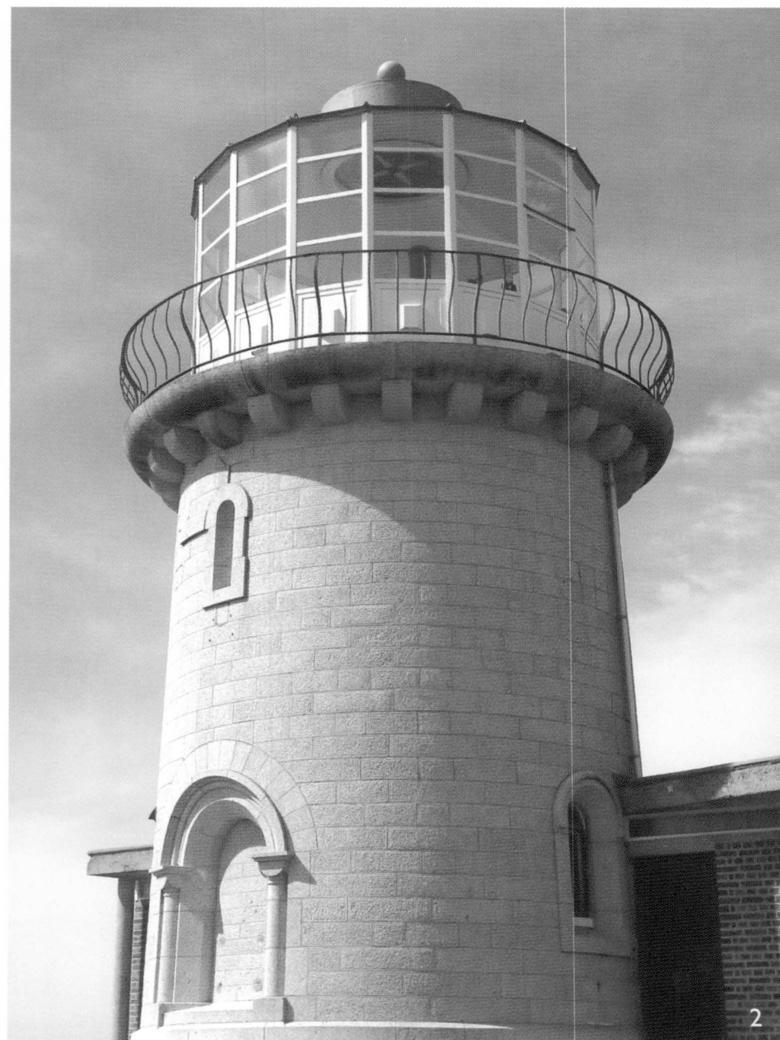
2

Our most well known towers are Eastbourne's two lighthouses. Belle Tout (shown above right) was constructed between 1832 and 1834 and designed by W. Hallett and J. Walker. This 40ft high tower of Aberdeen granite was built on the top of the cliffs and found to be useless in foggy weather. The lantern is a replacement carried out by the BBC for their film *The Life and Loves of a She-Devil* in 1986. The replacement lighthouse at sea level below Beachy Head is 141ft high and opened in 1902. It is constructed of 3,660 tons of interlocking Cornish granite blocks designed to withstand the considerable force of storm rollers and was manned until 1983 by three lighthouse keepers. It has recently been repainted following a local campaign "Save Our Stripes".

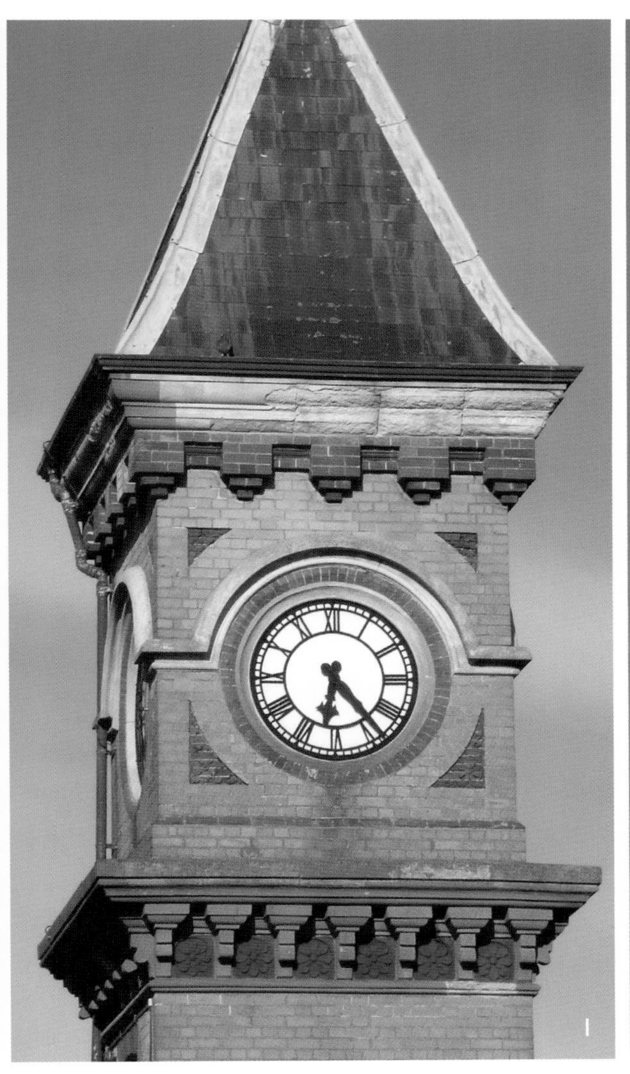

this page:
1. Eastbourne Railway Station
2. Holy Trinity Church, Trinity Trees
3. Kings Arms public house, Seaside

facing page:
1. Beachy Head Lighthouse
2. Belle Tout Lighthouse

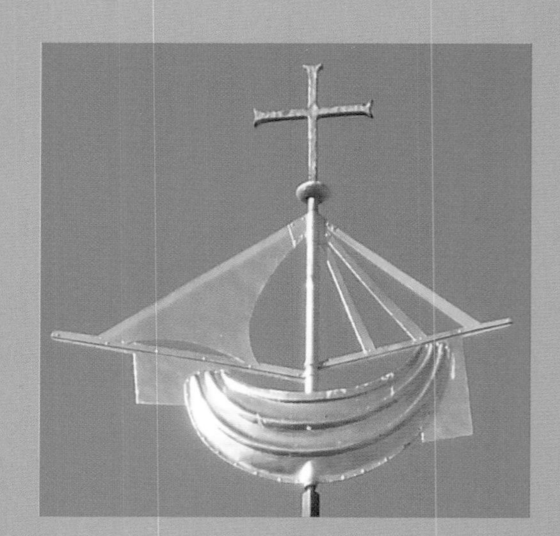

20

weathervanes

The ultimate piece of decoration, the weathervane has for centuries been reserved a special place in the hierarchy of architectural detail on buildings and gained an almost spiritual significance. They originated from pennants (or pennons) carried on lances with badges or crests used in battle. The weathervane picks up a thing which has become an obsession with the English, namely the weather – the one thing which humanity has not yet been able to control. The ornate pennon vane at the top of St Mary's parish church was erected in 1868 at the end of a major restoration of the church. The stencilled initials refer to Alexander Hurst and James Gorringe who were churchwardens at the time. Alexander Hurst was a churchwarden for 24 years between 1859 and 1883 and a member of the Hurst family who founded the Star Brewery in Eastbourne in 1777. These gentlemen served under the Reverend Thomas Pitman who was Vicar of Eastbourne for an astonishing 62 years from 1828 to 1890. Pitman was involved in all the major affairs of the town and witnessed the growth of the old village with 2,000 people when he arrived, to the fashionable resort with a population of 40,000 when he died. He is pictured at the opening of the Langney Outfall on page 4.

20

Another pennon vane dated 1842 that used to grace the Star Brewery was relocated to the Old Courthouse in Moatcroft Road when the brewery was sadly demolished in the early 1970s. It displays the founding date of the brewery by William Hurst in 1777 and also the initials of Harry Hurst whose family in Eastbourne went back to 1653 (shown above). Weathervanes have for generations been formed in various shapes and styles. The Railway Station vane of 1886 is very much in the Gothic tradition whereas the pennon on the corner of Arlington Road and Old Orchard Road of 1902 has adopted an Art Nouveau style. The striking fishing boat vane on the top of the spire of Our Lady of Ransom Church comes straight from a biblical scene on the Sea of Galilee. This church was opened in 1901 but the tower and spire were not added until 1912 so the vane must date from this time. The building was designed by Frederick A. Walters who was later the architect of Buckfast Abbey in Devon.

page 195: St Mary's Church, Church Street

this page:
1. The Old Courthouse, Moatcroft Road
2. Eastbourne Railway Station, Terminus Road

facing page:
1. Arlington Road
2. Our Lady of Ransom Church, Grange Road

1902
N
S
W
E
1
2

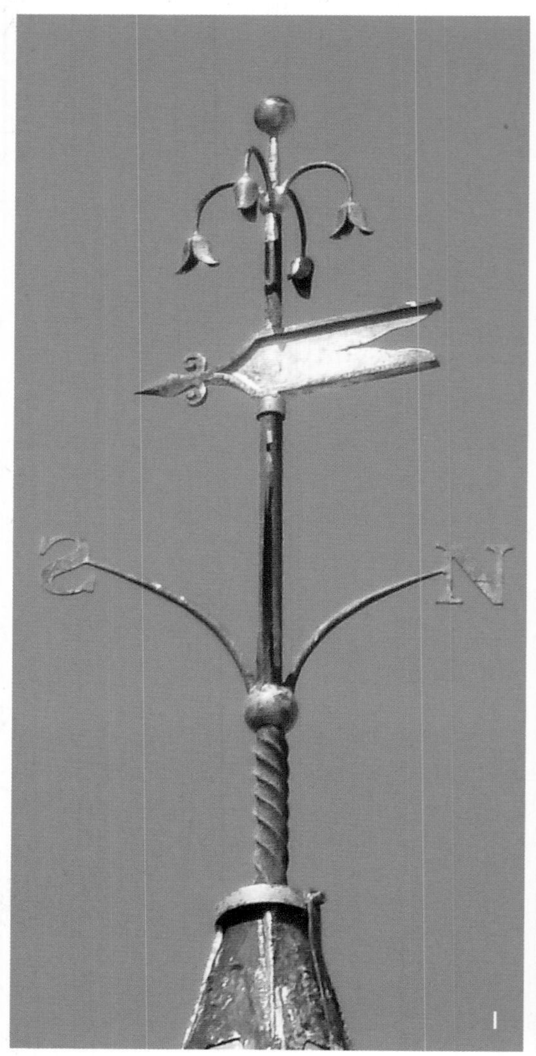

Weathercocks are to be found in various shapes and sizes. The cockerel on a church spire was a symbol of St Peter and also represented the parish priest as head of his flock. The latest Eastbourne weathercock is "Philip". He flies 175ft above the ground on St Saviour's Church in South Street and was made to commemorate the Diamond Jubilee of Queen Elizabeth II. This gilded bird is of beaten copper and contains a "diamond" glass eye with his cockscomb coiffured into a crown shape appropriate for the Diamond Jubilee. He looks small from the ground but actually measures 3ft long by 2ft high and in his tail is stencilled EIIR 2012. Further down the weathervane is a gilded crown which contains six glass "gemstones" representing each decade of the Queen's reign. He was made by blacksmith David Skinner of Blackboys and won a Sussex Heritage Trust Award in 2014. At the dedication service, the congregation on the ground sang "Pleasant are thy courts above" as they had in 1872 when the original weathercock was erected by the Vicar, only to be blown down in a violent gale soon after.

this page:
1. Leaf Hall, Seaside
2. All Saints Church, Grange Road
3. Silverdale Road

facing page: "Philip", St Saviour's Church, South Street

E II R
2012

afterword

The 'In Detail' Series

The starting point for the Salisbury book (see below) was a donation to the Salisbury Civic Society to commemorate a member's parents. The concept was developed by Richard Deane of that Society and the team that created the book included architects from a local practice within which the layout was conceived by Louise Rendell and Melanie Latham. Because Salisbury has many books depicting its cathedral it was decided that that building should not feature and overshadow the many other fine features in the town.

Drawing inspiration from that, the Peterborough Civic Society created their book to celebrate the 60th anniversary of the founding of their society and, coincidentally, the 125th anniversary of the Peterborough Photographic Society, and eight members of that society undertook the photography. This book did include the cathedral and also reached further into the countryside to include all of Peterborough's Unitary Authority area and just a few yards more!

In both cases the remit was to photograph what could be seen by the public without trespass.

Due to Melanie Latham's discussions with her father Edward Dickinson, Richard Deane was invited to come over from Salisbury to put the concept of creating an Eastbourne book to the Eastbourne Society and this discussion eventually led to particularly interested members of that society, especially Richard Crook, a local conservation architect, and Nicholas Howell, a professional graphic designer, to throw their weight behind an independent project. With further guidance coming from the Leader of Eastbourne Borough Council and the Compton Estate office, a steering group was formed and photographers were recruited including an involvement by Sussex Downs College. The same rules of photography were followed.

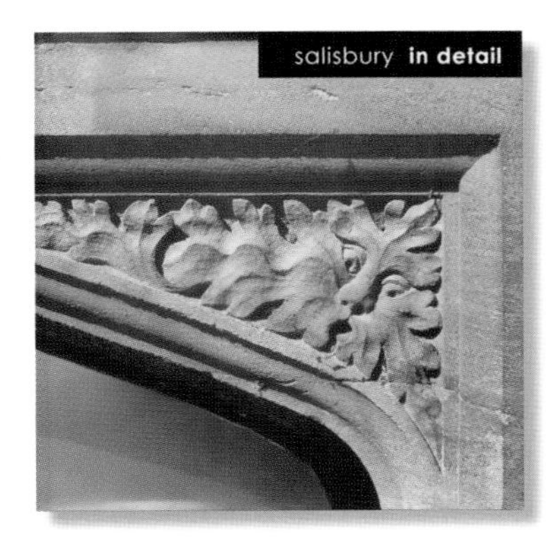

Published in 2009 by
Salisbury Civic Society
ISBN 978-0-9512100-1-7

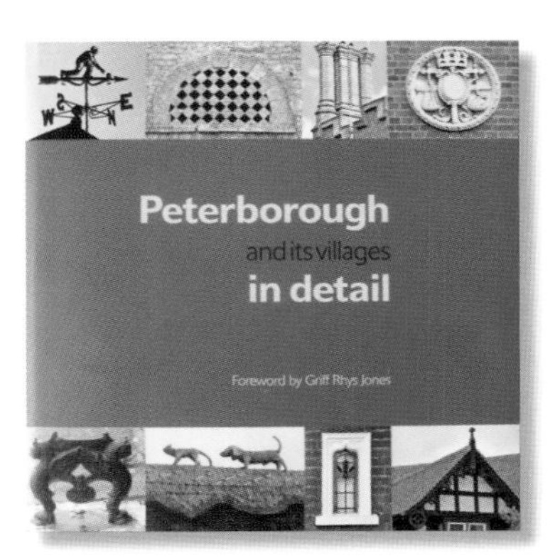

Published in 2012 by
Peterborough Civic Society
ISBN 978-1-907750-35-9

20 of the best

		page
01	Street furniture	9
02	Statues	19
03	Windows	37
04	Doors and porches	49
05	Door furniture	54
06	Brick and flintwork	61
07	Stonework	75
08	Ornate plasterwork	87
09	Decorative ironwork	101
10	Terracotta	113
11	Stained glass	117
12	Balconies, canopies and shelters	131
13	Decorative tiling and mosaics	136
14	Plaques	143
15	Clocks and sundials	154
16	Signage	159
17	Chimneys	169
18	Roof embellishments	174
19	Turrets and towers	188
20	Weathervanes	195